SOAPBOX

LAURENCE DAVIDSON

Also by Laurence Davidson

The Flow: A Hedonist's Guide to Buddhism,
Meditation and Travels in Thailand

Kill All the Butterflies

SOAPBOX – a novel

Google Translate was used for assistance with foreign language inclusions. No generative artificial intelligence (A.I. - LLM) was used in the writing or creation of Soapbox.

Soapbox is a work of fiction. Names, characters, businesses, places, events, and incidents either are products of the author's imagination or are used fictitiously. Any resemblance to actual persons, living or dead, or actual events is purely coincidental.

Published by Laurence Davidson & Flip Flops Off Press
Walla Walla, Washington
First Edition, July 2026
ISBN: 979-8-99-959533-1

Dedicated to

The Cybernauts

TABLE OF CONTENTS

PROLOGUE: THE SLINGSHOT

The office was empty except for Martin Reeves.

It didn't matter what his name was because a decade from that very moment one hundred million people would sit in some chair somewhere and feel the same thing. None of their names would matter either.

Martin stayed late. Again. The cleaning crew had come and gone. The partners had gone home to Atherton and Menlo Park and Los Gatos and their swimming pools and their certainties. The phones stopped ringing.

He had an IBM PS/2 running Mosaic 1.0 and a modem. He had a phone line and twenty minutes before he would drive home to his wife who would ask what kept him again and he would have no answer that made any sense.

The modem sang its dial-up aria. Discordant and mechanical. It was, in its day, the harsh screech of connectivity.

He launched Mosaic and typed in an address that a client emailed him.

And then… goldfish.

They were swimming in a tank somewhere in California, captured by a live feed camera, sent through a wire and reassembled pixel by pixel on his monitor in his office in Santa Clara at 11:17 on a Tuesday night. The fish didn't know he was watching. Nobody knew he was watching.

He sat back in his chair and looked at those goldfish swimming in their tank for a while.

What he was feeling didn't have a name yet. The vocabulary hadn't been invented because the experience hadn't existed long enough to require one. But if anyone pressed him, if his wife had been there, or his law school roommate, or his professors, or anyone who knew him enough to ask the right question, he might have said something like this:

Every idea I've ever come up with formed in my mind and stayed there until I opened my mouth or picked up a pen, wrote it down or typed it out. Then it traveled as far as my voice carried or a different form of communication reached its destination. That was the deal. That was the deal for every human being who ever lived. Your thoughts were yours... until they were the recipients.

And now...

He stared at the swimming fish.

"Well I'll be damned!" he said aloud to no one.

He'd had this feeling once before. Not personally, of course. Only by proxy, through history. Gutenberg.

Gutenberg had taught humanity how to speak for and to itself. The Reformation, the Enlightenment, the Scientific Revolution, all of it streamed out of that grand mechanical innovation like light through a prism.

It had taken a generation for people to understand what Gutenberg had done. Not the press or the printed books or that it made knowledge cheaper.

He redistributed the power of thought and knowledge and ripped the control of them from the Church, from the aristocracy, from the gatekeepers, and handed it to anyone who could read and write.

That had taken fifty years to fully ignite.

Now this! This... this was teaching humanity how to answer.

Martin Reeves thought, watching fish swim in a tank in a building he would never visit, watched simultaneously by strangers whose names he would never know and who would never know his... This was Gutenberg all over again *and this* was going to take about fifteen minutes.

Gutenberg needed a printing press and supplies. He needed a building, workers, capital and distribution networks. Months between thought and publication were the norm.

Goldfish needed a phone line and a computer and those were becoming commonplace.

Anyone could publish. Anyone could read. Anyone could respond. The medieval scholar at Oxford and the factory worker in Detroit and the philosophy student in New York and the retired teacher in Osaka and the programmer in Helsinki and the poet in Buenos Aires... all of them, all at once, in the same infinite room, with equal access to infinite conversations.

No editors or gatekeepers or Bishops deciding which ideas were permissible. Just minds talking to minds. Across every border and time zone and timing of birth that had previously made those conversations and the sharing of ideas impossible.

This is the Enlightenment, Martin thought. *With a modem.*

He didn't sleep much that night.

Nobody who witnessed the birth of the world wide web in 1993 did.

They were too busy imagining what the conversation was going to sound like when *everyone* logged in.

Nobody could imagine the rapid spread of the modern printing press. It was too vast a proposition.

Except for Martin's client.

PART I

SIGNAL

Zhen Sheng. San Francisco.

The alert pinged in his earbud at 4:47 PM. Over decades, he had learned there were no convenient moments when anomalies announced themselves.

Zhen Sheng walked to his enormous wall-mounted monitor and looked at what was scrolling in the systems monitoring report app. In thirty years of security work, he had developed an instinct for the difference between noise and signal, the digital equivalent of hearing a single wrong note in the middle of a symphony. Most people couldn't hear it. He could. It was, his employees at Argus Secure sometimes joked, his superpower. Not his patents. Not his systems architecture work. Not the threat modeling that made him the most respected security evangelist around the globe.

The alert itself was minor. A routine flag. The kind of thing a junior analyst would clear in forty seconds and forget by end of day. An anomalous weighted entropy spike in the third-party egress layer. It looked like a standard handshake delay, a momentary stutter in the API. But the telemetry showed the delay wasn't random. It was proportional. The system was holding back high-sentiment outrage packets for exactly 400 milliseconds, just long enough for the Prizmatik mobile client to refresh the user's bio-sync.

It wasn't a lag. It was a cocked hammer. He had been expecting something like this. Seeing the telemetry quantify it in cold numbers was something else entirely.

He copied the alert into a separate file, labeled it and added three notes that looked like paranoia to anyone else. It was simply due diligence for Zhen.

He closed his laptop, put on his jacket, secured his office with the custom biometric systems he'd developed back as an undergraduate student, and took the elevator down to the ground floor at 201 California Street, Argus Secure headquarters… into the late afternoon San Francisco light he never once took for granted.

The fog was rolling in mistily off the Pacific and through the Gate.

When he had arrived in The City from Beijing, young, electrified with possibility, carrying two suitcases and a head full of ideas that owned no outlet yet, the fog disoriented him. In Beijing the sky was either there, or it wasn't. In San Francisco the sky was a minute-by-minute negotiation with beauty. It gave and took light according to its own whims and you either made peace with that, or you left.

He made peace with those ever-changing moods within two weeks.

Zhen walked around the corner and crossed Battery Street to a parking garage along Halleck Alley, then down two floors to his vintage gun-metal grey Maserati Quatroporte III, started it and listened to its signature Italianate growl purr through the garage. It was a sound he never tired of. He drove west up California Street through the city with a practiced ease of someone who spent decades learning its rhythms and how to use a clutch. He was in Presidio Heights at 5:52PM.

The house wore Edwardian bones, updated long ago. From along Pacific Avenue, it announced nothing. Presidio Heights homes rarely did, which was one of the reasons he'd chosen the neighborhood. It was brooding, having been painted, prior to sale, a deep dark cacao that created a gravitational hole along the block. Inside it was something else entirely. High ceilings. Warm cream walls. Glowing hardwood floors that had absorbed thirty years of dinner parties and Wagner at considerable volume and the barefoot pacing of a man who did his best thinking in motion. The walls were covered in art he had collected from galleries and showings on five continents. Not prints nor reproductions but actual works from living hands, some of them by artists who sat at his table and shared meals and wine and absinthe and argued about things that mattered until the early hours of following mornings.

He changed out of his work clothes, made a double espresso on the La Pavoni machine that had taken him four months to calibrate to his satisfaction and took it into the atrium.

The atrium was one of his indulgences. The greenhouse addition off the kitchen offered heat and humidity, controlled and lit by full spectrum panels that cast light across everything growing inside. Four different heirloom tomatoes, cucumbers, three varieties of basil, rosemary, two kinds of thyme, Greek oregano, three varieties of mint, flat leaf parsley, French tarragon as well as an assortment of exotic specialties; lemongrass and a Kaffir Lime tree that was now six feet tall. Next to it was a Meyer lemon tree in a terracotta pot that didn't have any realistic expectation of thriving in San Francisco.

Zhen Sheng privately considered the sheer stubbornness of both he and the tree a beautiful, symbiotic relationship. Orchids of a dozen shapes and colors ran along the south wall. And tucked in the far east corner, tended with the same assessing eye and careful hand as everything else, two plants of an indica strain so deeply, glittering purple that visitors, if they were allowed into his sanctuary, sometimes stopped mid-conversation to stare.

He cultivated the strain himself over a decade. It was some of his better work. He'd put it against any of his other creations.

He stood in the humid air of the atrium, espresso in hand, and breathed deeply, slowly and methodically. The day fell away in layers. The anomaly didn't fall away. 400 milliseconds, repeated. That was the nature of signal. He would review it tonight with fresh eyes, and either confirm his instinct or discover he was wrong and sleep soundly either way.

He was rarely wrong.

He snipped rosemary. He snipped thyme. He stopped and smelled them completely. The information they carried was important and deserved his full attention.

The 48" Wolf range and hood with its ANSUL system was the first thing he'd upgraded when he bought the house. It was non-negotiable. Everything else could wait. The listening room, the fine trim that needed repair or replacement, the atrium... but the range was infrastructure. He built from the range outward.

A ribeye had been coming to room temperature since he arrived home. USDA Prime, dry aged for forty days, from Bryan's Meats on California Street. They knew him by name, knew his preferences and did not require instruction. He salted it from on high with Diamond Crystals and fresh ground black pepper, flipped it, then repeating, achieving a perfect dusting.

The heavy 15" cast iron skillet sat where it always sat. Above his preferred 20,000 BTU burner. He turned the heat to high and left it alone. Cooking properly requires discipline. The pan has to be smoking-metal hot before anything touches it or the whole enterprise goes compromised. He explained this once at a dinner party to a venture capitalist who made ungodly amounts of money on four investments. The VC had nodded with a polite incomprehension. He was a man who never accomplished anything in the kitchen other than toaster waffles and micro-zapped takeout in fifteen years.

While the pan heated, he prepped potatoes. Small red ones, par-boiled until just yielding, then onto another cast iron grill pan with avocado oil, rosemary and thyme from the atrium and four cloves of garlic crushed with the flat of his chef's knife then minced fine. The smell when the herbs hit the pan was, he always thought, one of the better smells available to human beings.

He had smelled extraordinary things in his life: the backstage of the Bayreuth Festspielhaus before a Ring Cycle performance, the scent of snow melt in Yosemite Valley in spring, hinoki wood of a world-renowned sushi counter in Tokyo's Ginza district, the combination of eucalyptus and ocean that hit you on the Presidio trails at dawn... fresh herbs on hot cast iron pan belonged in such exalted company.

Bok choi and Chinese broccoli steamed for exactly three minutes. The potatoes were done. He gave the ribeye one last look. The color, the surface moisture, and the way it relaxed into itself after its hour at room temperature and placed it in the skillet.

The hot iron sizzle it made was almost musical.

He did not touch it for four minutes. Another lesson learned early and never forgotten. The beef would tell him when it was ready to be turned. It always did. You just needed to resist the urge to interfere before it was time.

The temptation to interfere before a thing was ready. He thought about the anomaly while he waited.

At first, it appeared to be a repeating 400 millisecond delay buried deep in the outbound engagement architecture. Small enough to dismiss. Small enough that most engineers would dismiss it.

But the more he examined the timing clusters, the less uniform they became.

Some response chains reacted at exactly 400-milliseconds. Others drifted higher. 430. 480. 520. Not random variance either. The escalation patterns remained strangely consistent despite the shifting intervals. That was what bothered him.

Uniform systems produced uniform inefficiencies. This did not. Something was adapting. The latency behaved contextually, as if the architecture was adjusting pressure based on the user receiving it.

It was a small thing. A data pattern. A series of misconfigured endpoints. Probably what it looked like on the surface.

Probably.

He turned the ribeye and listened to the sizzle recalibrate. Tossed in sprigs of thyme, added a generous scoop of butter to the pan... good butter, European style, eighty-two percent fat... and began basting, tilting the pan and spooning the foaming butter over the meat with a motion that had become, over many years of repetition, automatic.

Ideally, the web is supposed to be what it looks like on the surface, flashed into his analysis.

He plated everything with the care he brought to all things. The ribeye rested properly, with its perfectly caramelized crust. The vegetables arranged without fussiness but not without thought. The potatoes creamy and fragrant. They were exactly what potatoes aspired to be.

He poured a glass of a 2005 Vega Sicilia Unico he'd been waiting for the right meal to open. This felt like the right evening, because there was an engineered delay to ponder.

The Presidio was dropping towards hard dusk, the eucalyptus trees fading to silhouette, the closest Golden Gate tower, perfectly erect, visible through a gap in the fog, the illuminated orange-red of it improbable against its grey robe.

He ate at his table alone thousands of times. He'd eaten at his table with musicians, painters, programmers, and philosophers and at least two people who subsequently become billionaires; one who went to prison. He preferred the table alone or with people who had something real to say. The middle ground... the polite dinners, the networking events disguised as meals... he'd exorcised those from his life somewhere in his forties. He didn't miss those.

From the listening room, The Dead. *Workingman's Dead,* because some evenings called for a particular American beauty rather than German cosmic tragedy. *Tomorrow comes trouble,* Robert Hunter wrote. *It could always go wrong.* Zhen Sheng loved that song for the plaintive, collective cry from humanity Garcia brought to the lyrics. You could hear twilight in the first notes as they floated out from the Rega Planar 10 to the McIntosh C8 preamplifier to the blue-metered McIntosh MC1000 monoblocks, delivering sonic purity out to Bowers & Wilkins 801 D4 Diamonds.

The Grateful Dead filled the house as he ate slowly. Adaptive latency and the potential impetus behind them was a problem. He let his mind drift, making connections between things that normally didn't talk to each other. It always worked that way. His best work came not from methodical analysis but from a purposefully unfocused mind, allowed to wander through seemingly unrelated data, finding unconnected points catching at the edges, until something lined up.

He poured another glass of Unico and went back to the atrium. He tended to all of his flora with careful movements. You grew something, you were responsible for it. That applied to all aspects of his life.

Then back to the listening room and his laptop, connected to everything, everywhere with his personal layered security build that would have made the D.O.D. envious. He took a breath and navigated to the encrypted drive he kept synced in three separate locations.

As he scrolled past folders housing Usenet threads from 1994, early forum posts and the first website he built, its HTML so simple and elegant it was moving to look at now. A photograph popped up; someone took at a conference in November 1992 showed twenty people in a room, all of them certain they were standing at the precipice of something magnificent. He was in the photograph. Third from the left. He was young and cocky.

He closed the window. Opened the anomaly file. Read his three notes.

Then he did something he rarely did at home in the evenings. He opened a browser and clicked the bookmark for Prizmatik and scrolled to the comments section of a news article. It didn't matter which one. They were all the same now and scrolled for five minutes observing everything. That was all he needed. Suspicions confirmed.

He closed the laptop. Sat in the listening room, now dark except for the glow of the blue meters and orange tubes. He remembered the one time he had been drastically wrong. Similar pattern. Different system. What it cost him those many years ago. He did not follow that thought further.

The Dead finished as the stylus clicked in the end groove. He put on King Crimson and settled back into the thick caramel leather Chesterfield. Thinking, until the last notes of Epitaph faded out.

He recalled Du Fu.

國破山河在
城春草木深

The nation is broken. Mountains and rivers remain.
In the city of spring, grasses and trees grow deep.

About the grief of witnessing something magnificent collapse from a distance.

He said it quietly to the empty room in Mandarin, finished his wine, and went to bed as the anomaly waited patiently.

OPTIMIZE

Nadia Farouk and Dr. Joel Becker. THE COGNITIVE ARCHITECTURE GROUP - Prizmatik Campus, Menlo Park.

Nadia Farouk's presentation slides were precise. She knew how to deliver strike points with clarity and force, compelling others to see what they had overlooked. Clean data visualization and exacting language were her forte.

ENGAGEMENT OPTIMIZATION — Q3 RESULTS

EMOTIONALLY ACTIVATED CONTENT: +34% YOY
AVERAGE SESSION DURATION: 4.7 HOURS
CONFLICT-ADJACENT INTERACTIONS: UP 2.1X
RETENTION AMONG HIGH-FREQUENCY USERS: 94.3%

"These trends are consistent with my projections." Her English carried faint architectural remnants of her Egyptian heritage. Not an accent exactly, more an inflection. She learned English by examining its structure in detail through Strunk & White and magazines rather than rote memorization. "Emotionally activated content continues to outperform neutral content by a factor of three point four. This is stable across all demographics and geographic markers."

Around the conference table were engineers, a product manager and two members of the Behavioral Research team. All of them looked at the slides with focused attention. They all understood that Nadia Farouk's meetings were not the kind you drifted through while doodling on a notepad.

Dr. Joel Becker was looking at the slides too.

He was taking notes. Not transcribing but annotating. Writing thoughts adjacent to what was being said rather than recording critical points. That bothered her. She could not stop noticing it. He always had the audacity to question her methods and findings.

"The conflict-adjacent number," he said. Not raising his hand. Just saying it into the room in his mild manner, which somehow didn't feel like an interruption. "2.1X — That's interactions that were initiated by conflict in the content, or interactions that themselves became conflictual?"

A beat. She couldn't stand his smug retorts, even though his observations consistently led to a level of data comprehension that she didn't see herself.

"Both," Nadia said. "The distinction, however, is difficult to isolate cleanly."

"But worth isolating..."

"Ideally, yes."

He wrote something in his notebook.

She clicked to the next slide.

The Prizmatik, Inc. campus in Menlo Park occupied forty-seven acres, which had been the last agricultural plot on the peninsula. Before that... if you went back far enough... oak and scrub woodland with tall golden grass bordering salt marshland. None of that appeared in the literature.

The literature spoke instead of thirteen LEED-certified buildings, two full-time cafeterias serving organic, locally sourced products, and a cautiously articulated philosophy of work-life balance.

Each building contained banks of ultra-modern sleeping pods, meditation cocoons and massage chairs. There was a private movie theater, game rooms outfitted with all the latest toys, and an outdoor kitchen designed for team-building barbecues. Retro Segway PTs stood ready for cross-campus movement. Electric charging stations lined every parking lot. They were always occupied. The buildings were clad with living walls of native plants while rooftop gardens regulated interior temperatures and air quality with quiet efficiency.

It was, by any corporate measure, a generous place to work.

The Behavioral Research division occupied the fourth floor of Building Seven. It had the best views of the campus and the hills of the peninsula to the west. That was not accidental. Nadia deliberately negotiated for that floor specifically when she built the division eight years ago. Environment shaped cognition. She knew this better than almost anyone alive and designed the team's environment accordingly.

Floor-to-ceiling windows allowed in maximal natural light. The furniture was ergonomic and unobtrusive. Digital whiteboards and interactive monitor panels covered entire walls. The custom coffee blends, of which there were twenty-one, were excellent. There were barista-level machines on every floor in every building.

The team did better work here than they would have done in the basement. She had the data to prove it.

The only data point she couldn't optimize was Joel's title. He was her boss and, in her calculations, for the simplest reason imaginable: he'd finished his Ph.D.

The room emptied. Everyone had more meetings. They always did.

Nadia closed her laptop and collected her notes, which she did in the same order she always did, a ritual that her therapist suggested, and that she maintained long after it stopped being effective. Like everything else in her world, it was structural.

Becker was still in the room.

He was standing at the window gazing down at Taleb Pond where the black swans lived. Seven of them glided atop the water with effortless majesty. They didn't have much choice. Their wings were clipped. His reflection was translucent upon the glass as he glanced up at the verdantly green hills beyond.

She picked up her laptop, then paused.

"Good questions in there," she said.

He didn't turn around immediately. Then, with the expression she was coming to recognize, the one that looked like agreement but wasn't, he said "Just grasping the full picture."

"The full picture is on the slides, Joel."

"Sure, most of it."

She looked at him for a moment. He looked back with his open face, that collegial directness, yet the power of his position evident. He appeared, she thought more than once, like someone's idea of what a behavioral scientist should look like. Guileless. Earnest. He wore the performance of it with an ease that made people in meetings overlook him. Nadia sometimes caught him staring at his reflection in the glass with an expression that wasn't academic curiosity, but a slow, quiet rot he was trying to drink away with expensive vintages of French and California cabernet sauvignon and pinot noir.

She underestimated him often. She consulted with him years ago because of the quality of his published papers and reputation. She was beginning to consider whether that had been entirely wise. Especially when the Board brought him in and elevated him as the CPO of her division because of one devastatingly accurate reassessment on a data string that she'd spent weeks on, creating such an obvious logic train that it was unassailable.

Dr. Becker took two minutes to dismantle it from every direction.

"The distinction you raised," she said. "Between initiated conflict and resultant conflict."

"Yes?"

"It's a real distinction. You're right that it matters."

He was good at waiting.

"Run the isolation analysis if you want to. Use the Q2 dataset. Let's see what you find."

Something moved in his expression. He caught the shift she always used when attempting to usurp his position to her benefit.

"I will. You know that though," he said.

She nodded once and left.

She stepped into the elevator and pressed the button for the ground floor. The doors closed with a soft hydraulic seal.

She replayed their exchange in the conference room, not for content but for structure. His interruptions were not random. They followed a pattern with the pressure applied at nodes of ambiguous or vague points. He wasn't challenging conclusions. He was isolating the mechanisms.

That was a problem.

Most people engaged with outcomes. Becker engaged with process. Process was where architecture lived. Process was where control either existed or didn't.

The elevator descended.

She had seen this before. In Boston and in Zurich, during two separate research panels that produced brilliant work and then collapsed under the weight of one person who insisted on asking the wrong kind of correct questions.

Not opposition, but friction.

Friction slowed systems and slow systems lost advantage.

She did a quick internal calculation without emotion, without narrative.

Could he be managed?

Possibly.

Could he be predicted?

Not reliably.

Could he be contained?

The elevator doors opened.

She stepped out into the lobby and made the decision the way she made all meaningful decisions... by removing the unnecessary variable from the equation.

Becker would continue to operate.

Just... not inside the parts of the system she built that mattered.

Joel stood at the window for another minute after she left.

The campus was beautiful in the late afternoon light. He thought so the first day he arrived and he still thought so now, which surprised him slightly. Familiarity corroded beauty. It hadn't done so at Prizmatik. Yet.

He stared at the living wall cladding the exterior of Building Six. Native plants, the campus literature extolled. He looked them up once, in his earliest days on campus. Some of them were native. Some were what landscape architects called *native-adjacent* – species that occurred naturally within a broadly defined regional ecosystem but not specifically here, not in this particular soil, not without significant irrigation, fertilization and maintenance.

Native-adjacent. Close enough to belong. Engineered to survive.

He examined his notebook. At the annotation he wrote during Nadia's presentation, next to the conflict-resultant metric:

> *Who and what benefits through isolation and conflict amplification?*

He closed his small black notebook and took the elevator down through Building Seven and out into the balmy Menlo Park afternoon.

He sat in his Tesla, the cabin silent and climate-controlled to a perfect 72 degrees. He didn't engage with the systems. He just sat, lost in the "Who benefits?" of it all.

He tried to reverse, but the car emitted a sharp, judgmental chime and refused to budge. A red graphic directive lit up the console: *DISCONNECT CHARGE CABLE TO DRIVE.*

He'd done it again. He exhaled, a long, weary and frustrated sigh, got out and unplugged it.

He hated the Tesla. Hated the way it presumed to manage his errors and hated even more that he provided it with so many errors to manage.

Not for the first time, he vowed to buy something vintage, mechanical, and analogue. Something that would let him ruin it without all the warning chimes and lane correction shimmy.

The isolation analysis would take him three days. He was already running a sympathetic version of it, for six weeks, for his own amusement, which Nadia didn't know, using a methodology that was technically on-point and very outside Nadia's toolbox.

What he found in his informal analysis and sandbox testing was clarifying. Which was, in every way, worse.

He pulled out of the Prizmatik campus onto the Menlo Park streets, past the living wall of native-adjacent plants, into the ordinary world.

Everything outside the gates behaved as if nothing changed.

PRIZMATIK SYSTEM LOG — 18:47:33.2

DAILY ENGAGEMENT SUMMARY

DATE: [CURRENT]

DAILY ACTIVE USERS: 2,400,000,000

AVERAGE SESSION DURATION: 4.7 HRS

CONTENT PERFORMANCE SUMMARY:

- OUTRAGE CONTENT: 847% AVG REACH MULTIPLIER
- NUANCED CONTENT: 23% AVG REACH MULTIPLIER
- CONFLICT-INITIATED INTERACTIONS: 1,400,000,000
- CONFLICT-RESULTANT INTERACTIONS: [ISOLATION PENDING]

BEHAVIORAL MODIFICATION INDEX: 8.9/10

NOTE: METRICS WITHIN OPTIMAL PARAMETERS

SYSTEM STATUS: OPTIMAL

RECOMMENDED ACTION: CONTINUE

SLCNMSTER69

Herbert Jones. A Basement. 3:17 AM.

Truth! Nobody wanted to hear it. Herbert Jones – Herbie to his friends, if he had any – knew that for years. Decades, really, if you counted early UseNet flame wars where he'd first laid out, at considerable length, why everyone else was wrong about everything from Pascal and COBOL to compiler designs to operating systems and motherboard architecture.

His positions should have been undeniable but was met with willful ignorance that confirmed everything he already suspected of the average person's relationship with facts. They didn't want facts. They wanted comfort. They wanted the warm milk of consensus spooned into their mouths by whatever ideological wet nurse with a following was trending that week.

SLCNMSTER69 was not that guy. He was the guy who threw the spoon away.

Now Prizmatik was in his sights and once again, nobody wanted to look where the evidence pointed.

On one computer, he opened his Winsock stack of old interfaces to feel like the world was still simpler. IRC, a Gopher client, Archie and FTP on one of his systems that still ran almost clean.

Everything was logged into a custom Lotus 1-2-3 spreadsheet system that was never built to handle the load of modern API data streams, except that beneath the Lotus interface sat nearly twenty thousand lines of custom code that translated modern network traffic into something the thirty-year-old spreadsheet could understand. Herbert built it over several weekends, one protocol at a time. Other people adapted to new software. Herbert adapted software to himself.

Then, on his newer machine (which in anyone else's world was archaic) he typed a thread he'd been building for four hours:

The Prizmatik Data Harvesting Operation: A Comprehensive Breakdown For Those With Enough Intellectual Capacity To Follow It!

He felt the familiar electric satisfaction of a sentence that said exactly what needed saying. He read it back. Added three exclamation points. Removed two. Left one. He had standards.

The basement was permanently populated by his muse – a life-sized figure of Han Solo with a light saber Herbert created out of a fluorescent tube he'd spray-painted sky blue. Herbert believed that Lucas had blown it, that Han deserved a light saber of his own, so he made that happen. It stopped working after its maiden test but that was beside the point.

Han was underlit by a lava lamp and the combined glow of three monitors. Two operational, one technically operational if you didn't mind the vertical green line bisecting the left third of the screen, which Herbert decided not to mind because the alternative was admitting he hadn't fixed it in eighteen months. The monitors sat on a cracked door, held together by duct tape, laid across two sawhorses. That was what Herbert called his workstation. Anyone else would have called it a cracked door laid across two sawhorses. Behind the monitors, arranged with the haphazard density of an archaeological dig, were the machines.

The Gateway 2000 tower. Three Compaq Presario boxes, each with their own inoperable issues he'd been meaning to address. Two Atari units in various states of disassembly, their innards spread across a towel he'd laid down with surgical intention six months ago and not looked at since. A stack of beige towers acquired from garage sales over more than a decade, their logos a museum of brand mortality – Packard Bell, AST Research – stacked in the corner as a monument to promising beginnings. Power supplies. Motherboards. Hard drives. Cooling fans. Unopened boxes of floppy disks. IDE cables coiled like snakes in a plastic bin. A soldering iron on a stand, its tip oxidized dark from the last time he'd used it, which had been before the current administration, possibly the one before that.

He was going to fix all of it. He had a plan.

On the floor beside and surrounding his chair (a high-backed gaming throne in aggressively red synthetic leather, which groaned in annoyance every time his obese frame sat down) were archaeological layers of recent sessions. An open bag of Cool Ranch Doritos. Piles of greasy, empty pizza boxes. A fast-food bag, still inhabited by a few french-fries that would never mold. A two-liter Mountain Dew, flat, one-quarter full. A paper plate bore a half-eaten microwave burrito consumed at some point. Herbert thought it was earlier that day. It was four days ago. Empty bags from a previous Cheetos, Doritos and potato chip binges and empty trays of every flavor of Oreo ever invented.

A stained towel that hadn't been washed in weeks reeked of sweat from pacing back and forth as Herbert's thoughts accelerated ahead of him. It hung over Han's available arm – the one with the blaster.

His HTC phone charged on a cable which ran under the door and up the stairs to the outlet in the hallway because the second basement outlet on that wall stopped working in 2009 and Herbert had not yet identified the right moment to address it. The only app that still worked on the phone was the phone and a third-party Tetris game.

Above him, the house was quiet. His mother had gone to bed at ten. He heard her move through her routines. The blaring television gone quiet, her walk to the bathroom, the creak of the third riser on the stairs, and then silence. She slept, as far as Herbert could tell, with the uncomplicated ease of someone whose conscience presented no obstacles to rest.

He envied her that. Some nights he could blissfully enter that state too, though only with the help of a heavy chug of NyQuil.

Not tonight. Tonight there was work to do.

The post was, he believed, some of his best work.

He'd been tracking Prizmatik's behavioral push notifications and response times for eleven months, feeding his findings with methodical patience into his Lotus 1-2-3 spreadsheet. He knew that the most important investigations were the ones the mainstream refused to touch. He had documentation. He had timestamps. He had a seventeen-tab browser situation that his lesser monitors could barely sustain and that constituted, in his estimation, a research infrastructure of considerable sophistication. His archaic spreadsheet software had served him well since 1994, so he saw no compelling reason to change.

Herbert's core finding was that Prizmatik was not merely a social media network platform. It was a behavioral modification engine, deliberately designed, in just one of the quiet corners of its interactivity, to identify emotional user volatility, to amplify their content, maximize conflict and harvest the resulting engagement data to refine the process. And of course, maintain usage to load up advertisements that would drive revenue.

It was, in Herbert's formulation, *"a dopamine casino rigged by psychologists who thought they were scientists but with the moral compass of a Vegas pit boss!"*

He was correct on the essential facts. He was also, in the presentation of those facts, a catastrophic liability to their credibility.

The thread ran to forty-seven paragraphs. It cited seventeen sources, nine of which were reliable, six of which were forums Herbert had posted in, and the rest of which were websites whose SSL certificates were lapsed, a detail he didn't consider even moderately relevant. It contained three genuine insights that a serious researcher would have found noteworthy, buried beneath language that included the phrases *"globalist data cartel," "feminazi engagement engineers,"* and *"sheeple won't read this but real smart people will."* He wrote with the absolute confidence of a man incapable of realizing his framing was the very reason no one listened.

It also contained, in paragraph thirty-one, a specific and accurate description of a data harvesting methodology that Dr. Joel Becker spent six weeks informally documenting in a sandboxed environment up the peninsula.

Herbert hit post at 3:23 AM and felt the release of it... the powerful satisfaction of having put something true into the world regardless of whether the world deserved it. He reached for the Mountain Dew. It was flat and tepid. He drank it anyway.

The engagement counter sat at zero.

He watched it.

Zero. And… still zero.

The anticipation of interaction raised his blood pressure another notch.

He opened another tab. Checked his follower count. Checked the follower count of a woman with a tech commentary channel who had four hundred thousand followers and whose analysis of Prizmatik's business model he read three months ago. He found it derivative and superficial and also suspiciously similar to points he'd made in a thread that received, at its peak, forty-one engagements, thirty-seven of which were his own replies.

Four hundred thousand followers.

He closed that tab with more force than the left mouse button required.

Upstairs, a floorboard creaked.

Herbert's shoulders contracted with Pavlovian cruelty. He minimized three windows of pixilated naked women with precision. It was, Herbert privately believed, one of his more marketable professional skills.

Silence.

Then the articulated shuffle of his mother's slippers. The bathroom. A flush. The squeak of the tap, then water running briefly. The shuffle back. The click of her bedroom door.

He waited thirty seconds. Restored his windows and reached for the Doritos. The bag was empty. He had not registered eating the rest of them.

He dropped the bag on the floor beside his vast dietary graveyard and turned back to the screens.

The engagement counter on his thread moved.

One.

He leaned forward. Someone had not only found the thread in the four minutes since he'd posted it, but engaged with it. He clicked through with the controlled urgency of a man trying not to seem as interested as he was.

The engagement was a reply from an account called TRUTH_ARTILLERY_88.

bro this is based af //the Prizmatik psych team literally admitted to this in a leaked doc two years ago // nobody covered it

Herbert stared at this, then typed:

"Source?"

He leaned back into his gaming throne, annoyed at its inanimate groan, and felt, despite everything, the egotistical warmth of being seen.

Someone somewhere was paying attention.

PRIZMATIK SYSTEM LOG — 03:23:41:.6

NEW POST DETECTED
USER: SLCNMSTR69
CONTENT CATEGORY: CONSPIRACY/PLATFORM CRITIQUE
OUTRAGE COEFFICIENT: 9.4/10
ACCURACY INDEX: LOW CONFIDENCE
REACH PROJECTION: STANDARD
EMOTIONAL ACTIVATION SCORE: 8.2/10
ENGAGEMENT POTENTIAL: HIGH
ACTION: PRIORITIZE DISTRIBUTION
AMPLIFICATION MULTIPLIER: 340%
NOTE: HIGH-FREQUENCY USER
NOTE: ENGAGEMENT PATTERN — OPTIMAL
NOTE: USER RETENTION PROBABILITY — 99.7%
USER RESPONSE PROFILE: NON-AWARE
SYSTEM STATUS: WITHIN OPERATIONAL PARAMETERS

CONTENT

Suwa. San Francisco.

The ring light had been on for four hours.

She no longer noticed ring lights any more than she noticed electrical outlets, gum on sidewalks, or gravity. At some point they'd stopped being merely things and become environmental conditions. The artificial sun around which her professional life orbited.

When she was twelve, her mother, upon witnessing her earnest performance as a newscaster in front of a mirror, purchased a cheap eight-inch model from an online marketplace that promised professional results and delivered something closer to tuberculosis. The light flickered at irregular intervals and cast a faint bluish tint across her skin that made her appear permanently exhausted and mildly haunted. It failed after five weeks of minimal use

Her current ring light was a Godox LEDP260C. A fourteen-month-old review post had pulled 1,847,916 views, unlocking a brand deal which generated revenue that matched most people's quarterly salary. She described it on camera as literally life-changing with conviction. All available units sold out across the globe within a week.

Suwa had not looked vaguely consumptive since.

The living room of her Mission District flat was optimized with the same attention to detail she brought to everything visible on camera. Four west-facing windows overlooked Shotwell Street and filled the apartment with afternoon light that, for approximately seventy minutes each day when it wasn't foggy, transformed ordinary human beings into slightly improved versions of themselves. Healthier. More rested. Marginally more attractive. San Francisco real estate was expensive because people paid extraordinary sums for small miracles.

Suwa paid willingly.

The matte-black wall behind her filming setup was decorated with floating shelves displaying a carefully curated collection of technology products, design awards, a couple of silk orchids, and books she would never read but frequently repositioned for optimal sizing and color blending depending on lighting and mood. Above them hung a neon sign glowing softly in electric blue.

미래

Future.

Most of her Western audience assumed it was decorative. Several asked if it was Japanese. One memorable commenter confidently informed everyone that it was Elvish.

Suwa stopped correcting them, letting them engage and drive her numbers up.

Today's filming session centered around the newest iPhone. It was positioned carefully on the table beneath enough lighting equipment to illuminate several blocks of 24th Street.

Unfortunately, Bisou was also on the table.

This violated filming table policy.

The policy had been established after an unfortunate incident involving dirty paws scurrying across the flat lay shoot that ruined a sponsored product layout. The email exchange afterward required three separate apologies and a partial refund.

Bisou appeared entirely unconcerned by any of this history.

He sat directly in front of the phone, a cloud of freshly groomed candy-pink fur, a purple bow secured between his ears and an expression suggesting that the entire production existed only at his convenience.

Suwa studied the composition.

Then she nudged him three inches to the left.

The framing improved immediately.

"You know exactly what you're doing, I'll give you that," she told him in Korean.

Bisou blinked slowly, his head cocked at a sixteen-degree angle. His silence carried the confidence of someone who knew he would never be required to explain himself.

The livestream began a few minutes later.

"Hello my beautiful people. Suwa here, bringing you the latest and greatest from around the world."

Her smile appeared automatically. It wasn't fake. Merely practiced.

People often misunderstood that distinction. Authenticity and rehearsal were not opposites. Every professional musician practiced scales. Every athlete trained fundamentals. Every successful creator eventually learned which version of themselves communicated most effectively.

"The camera system on this thing is incredible," she continued, rotating the phone beneath the lights. "Low-light performance is excellent. The optical stabilization is excellent. Color balancing is excellent. I know everybody online says everything is insane now, but this is actually much better than good. OK – I'll go there… it's insane!"

The viewer count climbed steadily to thirty-one thousand and twelve. Fifteen minutes later it was at thirty-six thousand, two hundred and ninety two. It passed forty-seven thousand within an hour.

They were the kind of healthy numbers that generated sponsorship renewals and invitations to product launches.

Everything was proceeding exactly as expected until Bisou stood up, stretched with theatrical exaggeration, and wandered directly into frame.

The chat instantly transformed into a digital riot.

"BISOU!!!"

"LOOK AT *HIM*!"

"HE KNOWS HOW TO MAKE ME HAPPY"

"HE'S SO FLUFFY"

"PINK PUPPY"

"I WOULD DIE FOR BISOU!"

Suwa glanced at the engagement graph.

She posted "he reviewed it first  "

The jump increase was immediate and dramatic.

She scooped him into her arms and continued to review the iPhone features while scratching beneath his chin as Bisou stared directly into the camera. The audience rewarded the decision enthusiastically.

The graph climbed higher as the angle steepened.

Interesting.

By the time the livestream ended, the review had performed well by every conventional measurement. Viewer retention exceeded expectations. Audience participation remained strong throughout the broadcast. Sponsorship metrics would satisfy everyone involved.

Yet an hour later, sitting at her desk with a mug of Jasmine tea growing cold beside the keyboard, Suwa found herself staring at numbers she did not particularly like.

Analytics were among the few things on the internet she trusted.

People lied. Brands lied. Comment sections existed primarily to demonstrate what civilization looked like when left unsupervised.

Metrics, however, possessed an unpleasant honesty.

She opened up her historical data. Then she delved deeper.

Patterns emerged.

The newest iPhone review performed well.

The photograph of Bisou sitting beside a coffee mug performed better.

A conference recap from Beijing performed well.

Bisou wearing a small yellow rain slicker performed better.

A detailed breakdown of the newest AMD processor architecture performed well.

Bisou asleep in a laundry basket performed significantly better.

She continued scrolling as the evidence accumulated with quiet persistence.

Years of expertise and research. Years spent learning enough about technology to influence purchasing decisions for hundreds of thousands of people.

And every single graph seemed determined to arrive at the same conclusion.

The dog was winning.

Suwa looked across the room.

Bisou was asleep on a blanket, one paw twitching occasionally as if chasing something important through a dream.

"You contribute nothing," she informed him.

Bisou remained asleep.

She refreshed the dashboard.

The numbers increased.

Refreshed again.

They increased again.

A small hollow feeling appeared somewhere beneath her ribs.

The idea of competing with a six-pound pink dog was ridiculous. What bothered her was something else.

For years she believed the transaction was primarily straightforward. People arrived because she crafted knowledge into something they found useful. She translated complexity into clarity. She tested products, filtered noise, and helped people make decisions.

Value created attention. That was the agreement, wasn't it?

The question arrived quietly enough that she almost missed it.

She stared at the screen.

Outside, the afternoon sunlight slid over the buildings across Shotwell Street. Pedestrians purposefully strolled the sidewalk carrying coffee cups, backpacks, their private universes invisible to everyone around them.

Inside, the dashboard continued climbing.

The dog slept.

The audience clicked.

The numbers rose.

And for the first time in years, Suwa found herself wondering whether expertise was the product at all.

She closed the analytics window before the thought could develop any further.

Some questions improved your life. Others simply made it harder to continue doing what already worked.

Bisou stretched, climbed into her lap without invitation, and settled there. Ownership of the situation had been established.

Suwa scratched behind his ears absentmindedly and stared out the window.

Below her, life on Shotwell Street continued exactly as it always did.

For reasons she couldn't quite explain, it felt different.

PRIZMATIK SYSTEM LOG — 16:47:22.3

USER: SUWA_TECH_OFFICIAL

CONTENT TYPE: ANIMAL/HUMOR

POST: "HE REVIEWED IT FIRST 🐕 🧮"

ENGAGEMENT VELOCITY: EXCEPTIONAL [↑]

ACTION TAKEN: NONE

NOTE: RECOMMEND CONTINUED OBSERVATION

SYSTEM STATUS: WITHIN OPERATIONAL PARAMETERS

FOUNDATIONS

Gabriel Lerner. East 87th Street, Manhattan

The brownstone had been in his family since 1948.

His grandfather Isaac Lerner, son of a tailor from Minsk, aspiring to a better life, arrived in New York with twenty-seven dollars sewn into the lining of his suitcase and a surname an Ellis Island clerk couldn't pronounce. He was renamed into someone else. Isaac bought the brownstone for forty-two thousand dollars the year he made partner at the immigration law firm he worked at.

Three floors. Pre-war Manhattan. Exposed brick and beamed ceilings that belonged to a more ambitious era. The windows looked out onto East 87th Street. It was not a glamorous neighborhood at the time, but times change.

Gabriel had grown up in that house. At seventeen, he left for M.I.T. and did not look back. His parents settled into the second floor when his grandmother died. His father followed two years after that. Quickly, efficiently, the way he'd done everything. He was a man who had been busy his entire life and saw no reason to linger through dying.

His mother was in Boca now with no plans to return.

The brownstone was his by default.

It was the kind of home that appeared in real estate listings with words like *generational* and *irreplaceable* and numbers Gabriel looked up once. He hadn't looked it up again. He believed sincerely, and argued publicly, that the concentration of generational wealth in the hands of the few was among the central pathologies of capitalism.

He heated his bagel from his weekly delivery from Bo's and let the contradiction sit there unresolved as it did every morning.

The kitchen was extraordinary in the way things outlast their purpose. His parents renovated it in the nineties, during a period of optimism and easy money that had not increased his mother's actual relationship with cooking. A Viking 36″ six-burner range no one ever used to capacity. A massive Subzero refrigerator that required a personal relationship with the service side of the company. The Meile dishwasher was so quiet he often checked to see if it was running.

What Gabriel used was the counter-top toaster oven and a Nespresso machine his mother left behind. He found the coffee almost adequate. Anyone with genuine interest in coffee found it inadequate with prejudice. He'd long since decided that propagating a genuine interest in coffee was not something to be concerned about.

He toasted a bagel but forgot to reset the dial on the toaster from waffle to bagel. He burned it because he was pondering a conundrum that felt more important than breakfast. He cursed as he tossed the blackened halves into the compactor, then toasted another one.

When the newly toasted bagel cooled, he slathered it with full fat cream cheese. Then he layered on the lox from Zabar's. Capers, then red onion, sliced thinner than most people bothered with. His grandfather sliced it thin. His father had sliced it thinner. Gabriel took this trend, without thinking about it, one step further. Some things don't require examination. They just require a sharper knife and a steady hand.

He ate standing alongside the kitchen island. Sitting down for breakfast alone felt, for reasons he never examined, like a concession. It gnawed at him each morning, reliably, alongside the coffee he found almost adequate, and he had come to think of that small unnamable dissatisfaction as simply part of breakfast.

His phone lit up on the counter. A notification from Prizmatik. He glanced at it, then looked away.

He'd read it later. Then: no. He picked it up, read two posts that had nothing to do with him but were somehow in his friend's notifications, put it in his pocket and went upstairs.

The third floor was his sanctuary.

Books covered opposing walls of the office in cherry-wood floor-to-ceiling shelves his grandfather's hands built. Philosophy, theology, computer science, psychology, political theory, history, mathematics, economics, comparative systems and Dummy books on subjects he didn't care to master. They were arranged by a system of his own devising that traced the arc of civilization from its first premises to what he considered its inevitable conclusion. His old M.I.T. dorm-mate at Burton-Conner House had spent twenty minutes searching for a copy of Rawls "A Theory of Justice" while Gabriel watched with amused patience. It was next to Aurelius (Meditations was lovingly abused and the spine cracked) and Havel.

On his grandfather's partner desk, which commanded the center of the room, were three laptops.

The first one was personal, for the inane engagement and management of daily life.

The next, air-gapped, ran a Linux distribution. He configured it during a single focused weekend two years ago. He sometimes thought about that weekend with affection.

The third ran an encrypted peer-to-peer OS The Static used for its meetings. That's what the group called itself for lack of anything better *and* a complete lack of poetic imagination or need.

Beside the laptops, legal pads dense with his handwriting: small, disciplined, the kind of script that looked typewritten until you were close enough to see the imperfections. Next to the desk sat an industrial-strength cross-shredder because Gabriel was someone who thought carefully about what shouldn't persist.

He opened the air-gapped laptop, transferred the previous session data from an encrypted USB drive, and reviewed The Static's accumulated notes. Fourteen months of twice-weekly sessions. Research he considered the most rigorous independent documentation of Prizmatik's behavioral architecture in existence.

It was still unpublished and a source of tension he was managing with mixed results.

The Static was a grand total of five members. Gabriel founded it, though he privately disliked the implication that anyone could truly lead a conversation.

Then there was Arthur Vance, the Ph.D. philosopher, who wanted to release everything to not only the web, but major international news agencies as well and trust the information to move freely. He fervently believed truth was buoyant enough to surface and surround their target, exposing corruption. Gabriel found this appealing in the abstract, and reckless in practice.

Maddy Winchell, a comparative divinity grad student, wanted coalition. The research as foundational for genuine movement building; a slow institutional pressure, the kind of change that actually lasts. Gabriel found this admirable in a way that exhausted him.

Kael, last name unknown, a tech dropout – because every school he attended was years behind him – wanted to break something. Anything. A week ago, at 2 AM, Gabriel would have found this understandable. Today, with too much below-mediocre coffee in his system, he found it more than understandable. He wanted to break something too.

And then there was the Ghost.

No one ever met the Ghost. The Ghost consistently interjected facts and data that were incredibly detailed and shocking.

He glanced at his legal pad. Centered, at the top of the page, underlined twice:

`Who inside P knows what we know? This wasn't a leak. It's too clean for that.`

For six weeks that question was the center of discussion. The correlated data had reached a point of specificity that was hard to explain without an inside source. Either someone actively leaked the data and the Ghost forwarded it all to them deliberately, or was, and this was the possibility that kept Gabriel on the third floor until the small hours, actively trying to be found.

He stared at the line for a while. Then he opened the encrypted channel and typed:

`We need to talk about the insider. Tonight. Mandatory attendance. The greatest cruelty of the` architecture isn't that it hates us; it's that it lacks any kind of optics analysis to see us as anything other than a profitable variable to be manipulated. To the system, the person who cleans the floor and the person who streaks it are only variables to be used. We are becoming caricatures of our own lives.

He sent it. Closed the laptop. He sat within the quiet of the third-floor sanctuary for a while.

He made more uninspired coffee, went out the front door, and sat on the stoop. The Upper East Side existed with the settled self-awareness of a neighborhood that had been significant for a long time and intended to remain so. A woman walked an expensive dog on an expensive leash. A grocery delivery arrived next door. Like all deliveries to those addresses, it was always on time. Two boys sat on the steps two doors down arguing about which Spider-Man was the best Spider-Man, an argument Gabriel owned opinions about but kept to himself.

He remembered a meeting. A meeting he sat in on. It was part of his job.

That was three years ago, yet he still returned to it the way you return to an exact moment you understood something about the world you couldn't unknow.

The project manager/product developer meeting introduced a behavioral nudge algorithm. The phrase "low-engagement intellectuals," was delivered without irony into a room that received it as marching orders… because it was a data set classification, and classifications are neutral.

As the senior engineer, he'd helped design the early engagement loops that led to that meeting. The code reduced the breadth of what users actually engaged with. Gradually. Incrementally. It reshaped the feed until the intellectuals engaged and then reacted emotionally.

The room listened with rapt attention.

Gabriel had sat very still and quiet, realizing what they were optimizing for, and the part of him that was tired of it and the direction the platform was obviously heading said, *"I can't do this anymore."*

He said that to himself as he rose and walked out without a word before the meeting was over.

That was it, at first. A quiet refusal, voiced somewhere below the level of applied language. He'd been listening to that voice for years. It wasn't separate from him. It was him, or a part of him; the part that skipped careful parsing and went straight to the weight of issues.

He knew the psychology of it. The way the mind partitions under sustained stress and duress, the self that manages and the self that feels, the gap between them filling up with something that has to go somewhere. He'd read all of it. Understanding the mechanism did not make the voice quieter.

He took some deep breaths as he glanced up and down the street, marveling as he always did, at the monied perfection that was both stoic and vulgar at the same time.

He picked up his phone. Opened Prizmatik. Read three posts.

Put it face-down on the step.

You opened it again, the voice said.

"I know."

You're going to have to stop.

"I know. But I want to understand how it works on me – past what I built."

He said that aloud. Nobody paid him any attention.

A pause. Then: *That's not the only reason.*

He didn't answer that, because there was no answer that he understood. Everything eluded his cognition on how the system was manipulating his feed.

Two doors down, the boys reached an impasse on Spider-Man. One took out his phone. The other watched him do it, then took out his. Gabriel watched them and felt something that was not quite despair and not quite pity and was maybe, underneath all the analysis, something simpler: grief for what a person gives away without knowing it.

I'm going to take them down, the voice said.

"I know you are," Gabriel said. "I am."

Then let's make sure we know what we're doing. And more importantly, why.

He turned the phone over in his hand. The screen locked. His lanky, mop-headed reflection with his thick glasses looked back from the dark glass, slightly distorted.

He put the phone in his pocket, stood and went inside.

He didn't look at it again. For twenty minutes.

WITNESS

Vira Tkachenko. Prizmatik Building Seven, Floor Four.

I have a system.

This is the first thing to learn when you have nothing else. Before money. Before language. Before anyone in this country knew my name. Or cared to learn. I have a system. You build one or don't survive.

My system at Prizmatik begins at the loading dock on the east side of Building Three at 11:00 PM. I badge in with the white card that opens every door on campus except the executive suites in Building One, and the server farms in basements of Building Seven and Two. I have never been told why my card doesn't open some doors. I have never asked. I already know why.

Some doors are kept closed. They're always closed to those who are not on mission.

I start in the lobby of Building Three. It is meant to make you feel. I know this because I have watched it work on people. Starry-eyed hires who come to orientation tours and investors who arrive in expensive cars with expensive shoes and walk in with postures of people who are impressive and impressed. The lobby makes them stop. A living wall. The light. The size of the atrium. It rises four floors and is filled during the day by people who believe they are building something important.

At night it is quiet.

I clean it.

I wipe a microfiber mop across glass-smooth floors in long strokes my supervisor Dolores taught me in first week. Dolores is from Mexico. She has been cleaning Prizmatik for eleven years. She taught me the floor product ratios, which surfaces mark, which ones don't, which managers notice and which ones never look at anything outside their heads. She taught me everything except what I actually need to know. I had to learn those things myself.

In Ukrainian we say: *Навчання – гірке, але плід його солодкий.* Learning is bitter, but its fruit is sweet.

My mother said this. She said many things. I stopped being able to hear her voice a few years ago, which is one of many small deaths that happens when you have been away from home for too long. You do not lose people at once. You lose them by pieces. First the voice. Then the smell. Then the way they hold their hands.

I still see her hands. They are on my mop handle.

Building Seven is last. I save it for last. It is one that requires the most attention. By the time I reach it I am fully inside my own thinking. It is when I do the best work. Not the cleaning.

Ms. Nadia Farouk's team, a Behavioral Research group, occupies the fourth floor. I have been cleaning that space for two years. I know those desks the way you know rooms of the house you grew up in. Which surfaces accumulate, which ones are always empty, what arrangement of things tells you about people who are not there.

Farouk's desk is always clear. Everything closed and put away, locked and shut off before she leaves. This tells me she is very careful. I have watched careful people my entire life. I know what they look like from the outside. They look like her. Disciplined.

Dr. Becker's desk is different.

He works late. I know this. Four nights every week I arrive on the fourth floor. The desk lamp is on and monitors haven't gone dark. He either left recently or simply forgot. A notebook is there. Sometimes. A black one. I don't read it. Unless it is open. His handwriting is small and sideways, and notebook is mostly closed. But the monitors.

He does not always shut off monitors.

I am not the obvious person who looks for things. I came for a job and was handed a mop, a white badge and a visa status that required me to be grateful. What I have learned here, I have learned the way I have learned everything since I was fifteen years old and understood that my survival depended on information.

You read what is in front of you. You remember it. You say nothing.

This is not espionage. This is Tuesday.

The first time I understood what Prizmatik actually was, it was a Thursday.

During my third week doing this work I was cleaning a corridor outside the main conference room on the fourth floor. It has glass walls that frost over by pushing a button. This is for sensitive presentations. That tells me something; that they frost glass even where everyone is theoretically and philosophically aligned. There are things not everyone is supposed to see.

The frosting has a gap at the bottom. About eight centimeters. I was on my hands and knees cleaning the baseboard, which required that position, and through the gap I could see the bottom third of the whiteboard.

Numbers. Charts. Data visualization that turns human behavior into a clean upward line.

`BEHAVIORAL MODIFICATION INDEX: 8.9/10`

I read it twice as I cleaned the baseboard. I thought about optics and invisibility. I moved on.

Before Odessa. Before Russia. Before men came. Before I understood cruelty men are capable of, I studied at Kharkiv Polytechnic. Information systems. I was there two years before everything stopped. I was not philosopher or computer scientist or behavioral researcher. I understood data structures and knew what a behavioral modification index was and I knew that 8.9 out of 10 was not a number that described a problem being managed. It described a system performing as intended.

I finished the baseboard and moved on to next office.

Мовчання – золото. Silence is gold.

My grandmother's version. My pimps' favorite tool. In between those two things is where I spend most of my life.

I need to tell you about those men. Not all of it. There is no version of it that belongs here, in this story, on this campus where people argue about content moderation over organic lunch. But some of it. Because otherwise you will not understand why I am the way I am, and why I know what I know.

There was a woman at an employment office. She told me about work in another country. Good money. Legal paperwork. She wore a kind face. I have since learned that a kind face is a professional asset, like any other. It can be cultivated. It can be deployed. It means nothing.

I will not describe what came after.

I will tell you what it left me with: I can read a room in under four seconds. I can identify the person in any group who holds real power regardless of hierarchy. I see exits before I need them. I know which rules are real and which are performed as appearance. I know that institutions run on labors of people whose names they don't learn. That invisibility is infrastructure. I carry all of this in my body.

Not as memory exactly, for memory implies something you choose to visit or not visit. It is more like bone. Like the way your spine knows how to hold you up without being asked.

When I finally escaped, and the how of that leaving is a terrifying story for a different time, involving a woman named Lesya who I will love for rest of my life for what she did, I received the kind of education that no university offers and no one would choose.

I could read any system. I could find any door. I understood as intuition how power works when that power believes it is not being observed.

I came to California because Lesya had a cousin in Sacramento. I came to Menlo Park because cleaning work paid more than anywhere else for someone like me. It was my first indication that something valuable was under protection.

The dental benefits mattered. So did location. So did fact that certain buildings collected more secrets than dust. The benefits were not whole story. Yes. My teeth needed work. Three of them. Damaged from violence. This is how I came to clean offices of people who are doing to minds what was once done to my body.

Tonight I am on fourth floor. Dr. Becker's monitors are on.

I clean around his desk first. I always clean around his desk first, because if he comes back for something... his phone, his jacket, his notebook he sometimes leaves... I need to have a reason to be there. I have learned to always have a reason for everything.

The first monitor is open on a spreadsheet. Columns of data I now recognize after nineteen months of education: sentiment scores, engagement metrics, amplification, the vocabulary of what they do. I have learned this language in way I learned English. Through total immersion, by reading everything visible and available, by refusing to let anything remain illegible.

But there is a document open on the other monitor. I can read it. It is a big monitor.

I move the vacuum. I circle. I am, from any camera angle in this room *(and I have mapped the camera angles in every room I clean)* simply cleaning near desk. Doing my job. Invisible.

The document is titled:

CONFLICT-INITIATED VS. CONFLICT-RESULTANT: ISOLATION ANALYSIS. FINDINGS ASSESSMENT.

I read it without being obvious. I read quickly. I always read quickly. In Kharkiv my professors called it a gift; later, other men found other uses for it.

What I read in Dr. Becker's document is not new information. Not to me. I have been assembling this picture from fragments. The frosted glass, the unlocked screens, the conversations that continue in the presence of cleaning crew because we are furniture. Nothing more.

I have a notebook of my own. It is not on any desk in any building on this campus. It is in my studio apartment in Redwood City in a place I will not describe, written in Ukrainian, in a personal shorthand that took years to develop and that I am confident no one else could read. The picture it contains is complete enough.

What Dr. Becker's document confirms is that someone else inside this building is seeing it too. That data I assembled from the outside, from below, from the position of my mop and a baseboard and trash that people throw away without thinking about who picks it up, matches what he has been finding with his access and his methodology and his PhD and his black notebook.

He knows what this is.

He doesn't know what to do about it.

But he knows who it comes from.

I have been thinking about what to do about it for six months.

I clean the monitor with a microfiber cloth. I wipe the frame, base, cable where it meets the desk. I am thorough. I am always thorough. It is, at this point, a thing in my life I can completely control.

I move on to next office.

There is a man named Volodya in my apartment building. He lives on the third floor. He is from Donetsk. He has been here for eight years and works in logistics for a freight company. On Sunday evenings he sits outside on the front steps smoking. I sit two steps down and he shares cigarettes and we speak Ukrainian to each other for an hour, which is only hour of my week when I do not have to translate myself.

He says: *Віра, ти занадто багато думаєш.* "Vira, you think too much."

I say "when has thinking ever been a problem?"

He laughs. He does not have an answer. Neither do I. Thinking kept me alive and thinking brought me here and thinking has shown me something in Building Seven that I do not yet know what to do with.

But I know this: I am the only person on campus with no career to protect. No sunk cost in promise of what that place is supposed to be.

I came here and was handed a white card and a mop.

I have been paying attention ever since.

Свідок. Witness. That is what they call someone in a legal proceeding. But in Ukrainian that word is older. It comes from *відати*. To know. A witness is not someone who only saw. A witness is someone who knows.

I *know*.

I have known for six months what is being constructed in that building and what it does and who it does it to.

I have been waiting to understand what a woman with a mop and a white badge and no stake in system can do with that knowledge.

I begin to understand.

I finish the fourth floor and take elevator down. I badge out at the loading dock at 5:49 AM into the early Menlo Park morning, which is cool and smells of eucalyptus and aromas of a place where nothing has been allowed to be imperfect.

I walk to my car. A 2003 Honda Civic with 243,602 miles and a heater that works only on the highest setting, which in Menlo Park is almost always too hot, so I drive with window down and heat on full because that is an exact compromise my car offers and I have made peace with it.

I am thinking about Dr. Becker's screen. Behavioral modification indexes and rage matrixes. About men who taught me, without meaning to, everything I needed to understand about what Prizmatik does.

Правда, my mother used to say. Not as a statement. As a question. As something you kept asking.

Truth.

I am going to find out what it takes, in this country, in this system, in this particular moment, to make *правда* matter.

I have survived much worse than this.

I turn onto El Camino Real and point the Civic north and roll the window down as dark air comes in cold and indifferent, and think:

I have always been the most dangerous person in any room that forgot I was there.

THE BLUE HOUR

Building Seven Lobby

Blue hour was a cheat.

Photographers called it magic. Influencers called it authenticity. Marketing departments called it peak engagement. In reality it was little more than a twenty-minute optical illusion where the sky became an expensive shade of blue that everyone pretended was profound.

Suwa loved it.

The pre-dawn sky beyond the glass walls of Building Seven glowed in gradients of deep sapphire. Menlo Park slept beneath a blanket of darkness while the first suggestion of morning gathered on the eastern horizon.

Her followers would never know she had been awake for almost twenty hours. Nor would they know she'd consumed enough caffeine to qualify as a controlled substance.

They would never know she spent the previous thirty minutes arguing with a sponsorship representative about whether the yoga mat should be rotated eleven or fifteen degrees clockwise.

What they would see was effortless serenity because *that* was the product.

The Level Four Creator Lounge was transformed into a sleek production studio. Ring lights. Soft boxes. Reflectors. Three camera angles. Two backup phones and a portable Breville espresso machine.

A collection of carefully curated wellness products were arranged on a reclaimed and refurbished teak table that cost more than most people's cars.

Suwa stood in front of the setup wearing charcoal-gray leggings and a matching sports bra emblazoned with her 미래 brand logo.

She looked and felt incredible and she knew it. She felt good to be that incredible, that desirable and exotic. The confidence that came from knowing exactly how she presented herself to the world was one of her life's pleasures.

Her public iPhone sat mounted on a motorized gimbal.

Her private Samsung rested on a nearby table out of sight as always.

The Samsung contained the real world.

The iPhone contained the profitable one.

"Okay people. Today we're focusing on intentional gratitude, conscious movement, and neural alignment."

Neural alignment sounds culty but it also aligns well and converts. She leaned into her presentation.

"We're going to learn about mindful optimization."

The livestream producer from the yoga company immediately sent a thumbs-up emoji.

Of course she did.

Nobody knew what mindful optimization meant. That was one of its greatest strengths.

The countdown timer appeared.

Thirty seconds.

Suwa took a long drink from a stainless-steel bottle marketed as molecularly structured hydration technology.

It was water. Expensive water. But still... water.

Twenty seconds.

She rolled her shoulders and popped her neck.

Fifteen.

Her smile appeared.

It wasn't a fake smile. Not entirely.

It was her professional smile, refined through years of audience testing and analytics feedback. A smile with excellent conversion metrics.

Ten.

Five.

Three.

Two.

One.

The red light appeared.

Suwa the planner, Suwa the promotional professional, became Suwa, the public figure.

"Good morning, all my beautiful humans!"

Twenty-two million followers. The number still made her happy. Not because she cared about all those people. That would have been exhausting. She cared that those people cared about her. Trusted her. Bought from her messaging. That was different.

"Welcome back to another Blue Hour session."

Comments exploded across the screen.

"Good morning from Singapore."

"Love you Suwa."

"I need this today."

"Your energy is everything. Suwa incredibly positive!"

The dopamine rush flooded her system. There it was – the hit. It was the same for every influencer on the planet.

She guided them through breathing exercises. Then a stretch sequence. Then a discussion about intentional stillness.

Halfway through, she accidentally knocked over a small ceramic planter.

The succulent bounced off the floor.

The livestream comments erupted.

"PLANT DOWN. OH NO!"

"SAVE THE PLANT."

Suwa laughed. A real laugh this time.

Within seconds engagement tripled. Her analytics panel reflected the surge. People apparently enjoyed watching wellness influencers screw up, just like a real person. The retention curve dipped precisely where she'd expected, then recovered after the plant incident almost exactly on schedule. Human imperfection still outperformed professional perfection. She thought about how to plan that into a future broadcast.

The producer sent another message.

LEAN INTO THE PLANT!

Suwa nearly laughed again.

Everything was content. Even accidents. Especially accidents.

By the time the stream ended, engagement exceeded projections by twenty-four percent.

The sponsor would be thrilled. The producer would be thrilled. The finance department would be thrilled. Most importantly, Suwa would be thrilled. By surpassing goals by that margin meant her bonus would be substantial.

The numbers mattered. Anyone who claimed otherwise was lying.

She ended the broadcast with her familiar sign-off: "Until next time my beautiful humans... stay present, stay grateful, and keep choosing your best self." She made a kiss to the camera one second before the red light cut.

She immediately opened the analytics dashboard. The graphs climbed beautifully. Shares. Comments. Retention. Conversions.

She enlarged the graph. The curves overlapped almost perfectly. That shouldn't happen. Travel content rose every summer morning. Pets followed their own rhythms. Cooking engagement climbed around holidays and weekends.

Different audiences. Different creators. Different emotional triggers.

Yet these climbed together, synchronized almost to the second, as though someone had reached under the hood and turned a single hidden dial. Whoever designed it understood attention the way journeymen mechanics understood engines.

"That's... elegant," she said aloud though there was no one there that heard her. For a fleeting moment she forgot to admire the results and found herself admiring the machinery that produced them.

Then a separate notification appeared.

TOP PERFORMING CONTENT THIS WEEK

Suwa glanced at the list.

Number One:

BISOU PLAYS FETCH

Number Two:

BISOU DISCOVERS PEANUT BUTTER

Number Three:

BISOU REFUSES TO LEAVE BED

Suwa stared at the rankings then burst out laughing.

Three professionally produced wellness campaigns. Two luxury retreat sponsorships. An exclusive interview with a Nobel Prize-winning neuroscientist.

All defeated by a fluffy dog.

"Traitor."

She pointed accusingly at a photo of Bisou.

The algorithm, however, clearly disagreed.

People loved dogs. The platform loved dogs. Advertisers loved dogs. Apparently civilization itself loved dogs.

Her Samsung vibrated.

The packet capture application remained open as rows of data scrolled silently.

She glanced at it without much interest. Mostly out of habit with a touch of curiosity. It was, she told herself, something to occupy the analytical side of her brain.

The Ohio Tax Hike thread appeared among the trending discussions.

Still growing. Still contentious. Still attracting enormous attention. Yet something looked strange.

She frowned.

Engagement velocity for her post dropped. Not gradually. Abruptly. A valve had been closed.

At the same time several unrelated categories experienced sudden growth.

Travel. Pets. Cooking. Lifestyle. Celebrity sightings. The charts rose together with suspicious synchronization.

For several seconds she stared at the dashboard.

That was odd.

She enlarged the graph. Studied it. Tilted her head.

"Huh."

A message arrived from that session's sponsor.

Fantastic session!!!

Can we discuss a dog collaboration?

Suwa laughed again.

Of course. Always the dog.

She closed the analytics window. Closed the packet capture. Closed the Ohio thread.

Outside the glass, dawn arrived.

The blue hour was gone. The illusion was over.

For another day, millions of people received exactly what they wanted. A prettier version of reality.

And Suwa received exactly what she wanted in return.

Attention.

The exchange felt fair enough.

She gathered her gear and headed toward the elevators.

Behind her, the analytics dashboard remained open on an unattended monitor.

The Ohio discussion slowing. Dog content continued accelerating. Travel videos surged. Cooking shorts surged. Some Hollywood-adjacent couple got an annulment a week after their lavish wedding.

Every indicator moved exactly where it was supposed to go.

No alarms sounded. No warnings appeared. Nothing looked wrong.

By then she was already in the elevator checking the contract for a luxury canine wellness campaign. The conversion rate for the offer was fantastic.

PART II

PRISM, PRIZM, PRIZMATIC, PRIZMATIK

Westwood, Los Angeles California. Then.

It began the way most things that change the world begin. With something small and innocuous.

Prizmatik started with a cork bulletin board. Four feet by three, mounted on the wall outside the computer sciences department's graduate lounge on the third floor of Boelter Hall at UCLA, where people posted a rotating selection of party announcements, sublease opportunities, lost pet flyers, alongside a philosophical discussion about the nature of consciousness that had been running in installment form since October. Everyone read the weekly index card. No one responded to it, because responding required either having the foresight to bring index cards and a pushpin with you, or finding the original poster in person, which required knowing who they were, which required that they put their name on it... which they hadn't.

David Weiss had turned the nagging engagement problem over and over with a slow persistence like a man sanding a large piece of driftwood into an octopus. It was a problem of the purest sort, as it was insoluble and therefore demanded nothing from him but his own quiet observation.

It was a Monday afternoon when an idea struck — a sudden flash that flickered between other thoughts yet refused to be extinguished. It stayed with him through three restless nights, growing more ridiculous with every hour, a persistent haunting that ignored the protests of his better judgment. He spent those seventy-two hours in a fever of internal litigation, attempting to dismantle the thing through reason, trying to talk his soul out of a deal his brain already inked. But the concept sat there, patient and immovable, refusing to abdicate to common sense.

The anonymity of those messages on the board bothered him. It was a failure of reach. Every bulletin board, every flyer taped to a lamppost, and every notice stapled to construction-site plywood and vacant walls since communities first had walls, demanded a physical coincidence. You needed to be there, standing there, at the exact moment the information existed. The message lived and died within the narrow arc of whoever happened to walk past.

Then the idea hit.

Not as a slow accumulation of logic arriving at a conclusion. It arrived with a flash, as he saw everything at once. He worried that such an easy solution would disappear. Except it wasn't gone. It was still there at two in the morning and again at four. It was still there over bad cafeteria coffee and grainy scrambled eggs the next day, and again that night, patient and faintly glowing, like an ember that decided not to die.

He spent three days in internal litigation. He was prosecutor, judge, and the increasingly unconvincing defense. He brought charges: *it's too simple, someone's already done it, the architecture alone would take…* The idea didn't argue back. It just sat there, calm and immovable, and let him exhaust himself against it.

By the third night he knew he'd lost the case.

He rose, went to his desk, sat down, and started writing code.

David was architecture and the elegant back-end logic guru. In the assessment of people who understood such things, his designs were some of the most brilliant infrastructure of its generation. He envisioned systems. He saw connections between things before he saw the things themselves. He could hold an entire network topology in his head the way a composer holds a symphony. Not note by note but as a complete reflexive living structure, every dependency and relationship mapped, weighted, and connected.

Marcus Teller was not a programmer in any serious sense. He could write functional code that worked as required by his course assignments. Barely. What Marcus possessed was rarer: he understood at an innate level what people wanted to feel. Not what they said they wanted, but their subterranean hungers they didn't admit to anyone, let alone themselves. He'd grown up in Long Beach and then Culver City, where at the local community college he was the kid who read Borges and the Iliad for pleasure and could walk into any room and know within thirty seconds exactly how the personality of the place worked.

The transfer to UCLA, and the scholarship came as a gift from one of his instructors at the Community College who saw him for what he was.

David and Marcus lived together for two years in a two-bedroom apartment on Gayley Avenue that perpetually smelled of Tom Yum, curry, lime and lemongrass from the Thai place downstairs – a smell that became the envelope of late nights, building sense memory tied to exhaustion.

They got along well because they occupied completely different worlds. David seemed to emerge from the womb fully formed for the world of tech while Marcus, when asked by his counsellor what major he was declaring, merely shrugged. She re-studied his recommendation letter, then assigned him psychology. He accepted it without objection.

David showed Marcus the code in March. It was rough. Merely a functional skeleton, no interface to speak of, running locally on his desktop in his office in Boelter Hall. It did one thing: it let you post something, attach your name and reply email, and have it visible to everyone else who ever logged in with their ucla.edu email address. It also allowed readers to respond in a reply field that would automatically capture and display their name and email.

That was the entire thing.

Marcus sat looking at the code and rudimentary interface for a long time, the glow of the monitor reflecting in his eyes. Finally, he spoke. "What if you could find what you'd lost?"

David looked at him, his mind still buried in logic gates and architecture.

"Not lost pet flyers or my Ray-Bans went missing," Marcus continued, his voice taking on an intensity that David understood and saw the moment it began. It was usually something to pay attention to. "People. What if you logged in and the system showed you every lost family member or friend from your past? Everyone from your grade school or your high school who was out there, somewhere, wandering the same wires? What if your seventh-grade best friend was at Cal-Poly and you had no idea that they were even close by—and then, suddenly, you did?"

David was quiet. He wasn't thinking about the friends; he was watching the way the information would have to flow, the way the tables would need to marry and pivot. He ran the structure through his head as he envisioned infrastructure.

"The data model for that," David said, "is straightforward. Of course we couldn't limit the domain access at that point."

Marcus saw implication and smiled. It was a small, knowing expression. "I know. That's why I brought it up."

They called it Prizmatik.

It took a while to get there.

First, it was Prism. Clean. Accurate. Descriptive of the thing Marcus envisioned. The idea that the platform didn't create connections so much as reveal them, the way a prism reveals the colors already hidden in white light. David liked Prism.

Then, because neither of them was cool and both of them were acutely aware of that, and because someone else was already squatting on the prism.com domain, they swapped the S for a Z. Prizm. They looked at it on screen for a day before agreeing, without much discussion, that it looked like a trendy snowboard brand.

Marcus pushed further. Prizmatic. Prizmatick. He liked the implied automation. Finally, he dropped the C, landing hard on the K, a crisp mechanical click. Prizmatik.

David found the typographic rebellion amusing. Marcus considered it essential.

"Prisms don't create light," Marcus explained, with patience. "They reveal what's already there. All the colors were always in the white light. The prism just shows them to you."

David thought it was a perfectly reasonable and defendable metaphor.

Marcus said, "it's not a metaphor. It's a mission statement."

They launched Prizmatik in April on the internal network, a walled playground restricted to staff and students with the *ucla.edu* address as proof of concept. It was a single page with a clean, spare interface that Marcus designed and coded over one sleepless weekend fueled by the cloying orange sweetness of iced Thai milk tea. He knew first impressions were critical. He knew college students had been beaten down by years of utilitarian design. They expected things to be ugly and difficult.

Prizmatik was effortless and easy on the eyes. That was intentional.

An early user, a pre-med senior, watched his browser stall. The spinning indicator hung there, a digital taunt lasting just long enough to irritate him into hitting refresh. When the page reloaded, his world shifted. The list transformed into something he thought was amazing.

Names he hadn't seen in years, pulled from the ether. He had not searched for them. Yet there they were. His freshman roommate. The nerdy but cute girl from his sophomore anatomy class who would tap her pencil in a syncopated rhythm. He could still hear the cadence if he closed his eyes. Then, his cousin in Sacramento. They hadn't spoken since a funeral they both preferred to leave in the dark. His cousin was alumni. David's algorithm found him.

Across campus, in a windowless office that smelled of ozone and warm circuits, the lead systems administrator stared at a terminal that refused to stabilize. Load spikes were hitting in irregular waves, suggesting not growth but sudden, rhythmic surges. The network developed a pulse of its own and was now testing its elasticity. He scratched absently at the side of his head, reached for the phone, hesitated, and watched another spike climb, flatten, then double.

Marcus stood behind David's chair. They were watching the live user map bloom across the screen, points of light multiplying and clustering, linking in real time, growing exponentially. Marcus didn't say anything. The pattern was obvious. It was a digital stampede.

Within two weeks four hundred UCLA students signed up for accounts. The registered user accounts hit one thousand when Melissa Jürgens signed up. Neither remembered much about the celebration except the rancid taste of cheap tequila the next morning and how their friends took endless pleasure in recalling their adventures in UCLA's Inverted Fountain under a full moon sky.

Within six weeks, twenty-two hundred and forty-six. The department's server infrastructure, which had been built for light load journal-based research, not the continuous demands of a social platform, began making noises. It was the sound of hardware being asked to carry weight it was not designed for.

On a Tuesday morning, Gerald, the systems administrator, stood in the doorway of David's office. His expression hovered somewhere between the begrudging respect of an engineer and the cold fury of a man whose house was being burned down.

Gerald had a pointed conversation with David. "I don't know what you've built... but it appears to hate RAM and has declared war on my servers."

David went to discuss the issue with the department chair who directed him to the tech licensing division of the computer sciences department. That conversation introduced him to what his concept might be worth. He found the information simultaneously electrifying and faintly nauseating. He discovered the stray dog he'd brought home was actually a wolf, and that the wolf was made of gold.

Marcus had been studying the behavior logs for most of the afternoon, not line by line the way David would, but in aggregates... heat maps, session clusters, and the patterns that emerged once the user base crossed a threshold. There was something there.

He leaned back, ran a hand through his unkempt hair, then leaned forward again, isolating a segment of users who joined within the same seventy-two-hour window. Their activity curves didn't match the baseline. They spiked faster, stayed longer, and returned sooner. This was not just engagement. There was attachment.

"David," he said, without turning. "Come look at this."

David rolled his chair over with a spin, already scanning the screen before he stopped moving.

"This cluster," Marcus said, tapping the CRT. "These users aren't just connecting. They're locking in. Look at this return frequency. Look at the session durations. This is where it tips."

David didn't respond immediately. He zoomed out, then in, then out again, dragging the time window wider.

"It's not tipping," Marcus continued, a little sharper now. "It's accelerating. This is what we talked about. This is the point where it becomes self-sustaining."

David studied the display for a moment. "No… it's not," he said, his eyes never leaving the screen.

Marcus turned. "No?"

David tapped the screen once, then dragged a second dataset alongside the first.

"It's a cohort artifact. Look at the date span from that start point."

Marcus stared at the two overlays and nodded. For a brief moment, he didn't say anything but the feeling lingered.

"These users joined at the start of midterms," David pointed to a cluster of high-activity accounts. "They are experiencing limited social bandwidth, triggering higher online interaction. It'll normalize in two weeks when exams are done."

Marcus felt his certainty shift beneath him. David was usually right in these matters. But he held on for a moment and tilted just enough as he thought about the potential of a million avenues of interaction that might transpire.

"You're sure?"

David was already in motion and turned back to his own screen.

"Yep."

Marcus sat there a moment longer, looking at the curves that felt, ten seconds earlier, like revelation. They still looked and felt different.

He minimized the window.

He opened the live user map instead.

People were not using Prizmatik the way David and Marcus envisioned, then designed.

David built an electronic bulletin board. An elegant networked system. A bulletin board for the modern age that knew who you were and who you knew and could therefore show you things relevant to your actual life rather than the life of whoever happened to walk past a cork board or a flyer on a pole.

But people were not using it primarily to exchange information. Instead, they were finding each other.

Dorm mates from freshman year. High school friends separated by distance and drift. Former lovers cautiously circled back into contact at two in the morning. Study groups formed organically across campus. Weekend trips materialized out of nowhere. Parties. Road trips. Coachella plans. Reunions.

It was the impromptu BBQ's outside Rieber Hall that revealed what Prizmatik was becoming. People were checking in. They were checking in often. The platform was becoming a tool for communication rather than simply a place to post information.

A woman in the School of Nursing who had grown up in Fresno found her best friend from childhood. They were inseparable once but lost each other during a move in sixth grade. Her friend was now attending film school at UCLA, living less than a mile away on campus.

For years there wasn't a mechanism capable of returning them to one another. No directory except for a telephone book which was almost irrelevant the day after they were printed. No searchable database existed.

But there was the occasional memory, arriving at irregular moments: *I wonder what happened to her?*

They met for coffee and cried immediately. Not from grief over the lost years, but from the shock of recovery and reignition.

Marcus heard about it through a mutual friend. That story stuck with him for a long time.

Classmates.com had been wheezing along since 1995. It was a vast, sterile archive of registered high schools and graduation years; a digital graveyard of maiden names, yearbook photos, and tepid nostalgia. But it never solved the friction arising from either missing or enabling connections.

Users could list themselves but actually reaching out required subscriptions, paywalls, and a series of digital hurdles that made human curiosity feel bureaucratic. By the time most users navigated the process, the impulse to reconnect often withered, leaving classmates.com heading to the digital graveyard.

Marcus ran across campus to David's office and entered without knocking, panting. A manic grin plastered on his face as he burst into the room.

"We need to integrate the Classmates database," he blurted out as he caught his breath.

David didn't look up as he kept working on a necessary structural expansion. "We don't need their data. We'll build our own. It'll take two or three years for meaningful engagement. Eighteen months if we move quicker and get input from the user base."

"But it will be ours," David replied.

Marcus smiled. "Only if we're first."

Marcus leaned over the desk, invading David's field of concentration, forcing eye contact. "Hear me out. Classmates helps people remember each other, right? Prizmatik is going to help them reconnect. We *need* their database, David. We don't have eighteen months. We have this moment, and that moment is right now. It won't be '*now*' tomorrow."

David went silent. Marcus had learned the art of arguing with him; you didn't overpower David by being louder or persistent. You won by presenting the truth so undeniable and defensible that David arrived at your conclusion before you finished making the argument.

David nodded. "The integration... it's going to be ugly. Their data architecture is a disaster. I've seen the schema underlying their foundation. It's childish."

Marcus simply looked at him, steady and unblinking, a slight grin rising at the corners of his mouth.

"It'll take four months. Minimum."

"It's critical," Marcus said.

"Fine." David sighed, already feeling the kind of soul-weariness that comes from realizing you'd agreed to rebuild Hoover Dam with a garden spade.

But Prizmatik existed to solve friction. That was the core concept of the system David designed: the belief that friction was the enemy of connection; his architecture was built to eliminate it.

It took Marcus two phone calls and a midnight flight to Seattle to get Classmates to agree. The speed of the deal surprised David until Marcus explained the frank math of it:

"Classmates is a body with no pulse; we're a heartbeat looking for a body."

It took David five months to sweat out the integration, writing an API from scratch so elegant and so clean that the debugger had nothing to say.

They launched the Classmates integration in November.

Zhen Sheng was sitting at his desk in his office at the company he founded five years earlier. Argus Secure was a consultancy in downtown San Francisco, already profitable, respected, and not quite interesting enough for the part of his brain that ran hottest.

His assistant Winnie Chan forwarded him a link via email.

Have you seen this? My college roommate just found her brother on it. He's been in Japan for six years. She didn't have his address.

He hovered over the link for a moment, before clicking it.

He was on the Prizmatik website for under four minutes before he understood exactly what it was. He was on it for eleven minutes before he found Dr. Wei Chen, his closest friend from Beijing University, before he'd left for America. Wei Chen had gone on to graduate school in computer science at the University of Toronto, specializing in statistical language modeling, a niche that very few people believed would ever matter outside academia.

At first, they exchanged letters, then fewer letters, then none at all as the erosion of distance and busyness and the assumption that there would be time later, slowly owned their horizon then shut the door.

Wei Chen had registered with his Beijing University email address, his Toronto email address, and his current address at the post-graduate research institute in Waterloo, Ontario, where he'd been for four years.

Waterloo. A direct flight from San Francisco.

Zhen Sheng sat at his desk for almost two hours after he found Wei. He doodled on a legal pad as ideas spun. Then he began making notes. He picked up the phone. An actual phone. The landline on his desk and called the number listed on Wei's profile.

Dr. Wei Chen answered on the third ring.

They talked for two hours and forty minutes. Zhen Sheng's assistant knocked twice. He waved her off without looking up. His native Mandarin language came back immediately as they caught up.

When he hung up, he sat for a while, quietly. He reopened his browser, went back to Prizmatik, reviewing every page while taking notes as he browsed. He spent an hour reading everything he could find about the company from every source he had access to: the founders, the architecture, the UCLA origin, the corporate filing, the Classmates integration, the server infrastructure and the growth curve from a sample prospectus that shouldn't have been accessible.

He'd built companies. He'd seen promising companies. He'd seen overhyped companies. He'd seen companies with elegant technology and no market. He'd seen companies with mediocre technology and extraordinary timing. He'd consulted with companies and assessed their systems and understood, at the technical, financial and social levels that Prizmatik's growth curve was not business-as-usual.

What he found was very different from the sustainable growth curve of most start-ups vs. what a standard growth curve looked like in traditional business-building context. He knew the difference between a curve, a hockey stick, and an anomaly.

He leaned back from his monitor and stopped taking notes.

This was not a growth curve or a hockey stick. This was proof of concept for something that had never existed before, an anomaly.

He found it viscerally exhilarating.

He called Marcus Teller's number. It was listed publicly on the Prizmatik press contact page. That told Zhen Sheng two things: Marcus Teller had not yet learned caution, and Prizmatik had not yet scaled the way it needed to be.

Marcus answered on the second ring.

The meeting was at a coffee shop on Westwood Boulevard, two blocks from campus, because David and Marcus didn't have an office and that day their apartment was permeated by galangal and Massaman curry, and they thought it might be too much to be taken seriously.

Zhen Sheng arrived ten minutes early. He always arrived early for meetings. It gave him time to read the room before anyone knew he was there.

Marcus arrived exactly on time, which told Zhen Sheng something.

David arrived four minutes late, slightly out of breath, carrying a laptop bag dangerously close to structural failure while wearing an expression of someone who'd been thinking about something else until approximately ninety seconds prior.

That told Zhen something too.

He bought them coffee and sandwiches. He asked David to explain Prizmatik's architecture in detail. He asked Marcus to explain their vision and what they thought the platform could become.

After two hours of discussions, he shifted and asked them both, separately, what they thought the other one's greatest weakness was. The question produced a silence and then, from both of them, answers that were honest without being cruel.

That was the moment Zhen Sheng decided.

He told them he wanted to invest. He told them the number. He watched their reacions.

David's face could only be described as having a vivid hue of disbelief; then a microsecond of recalibration he quickly calibrated to the potential of a very different start-up campaign.

Marcus's face did something different. It went very still as he stared at Zhen. Quietly, he asked, "What do you want from it?"

Zhen Sheng looked at him.

"I want to see what it becomes."

Marcus held his gaze, then nodded once.

Marcus and David looked at each other, then back to Mr. Sheng, and said, almost in unison, "Okay."

The Westwood meeting sparked the pivot. In the spring of 2004, UCLA was still Prizmatik's lung, but breathing was becoming difficult. Zhen Sheng had watched David and Marcus across the table in that cramped coffee shop and understood, before they did, that their creation was about to outgrow the university's server racks. He also understood what that would mean for both of them.

Marcus was six weeks from completing his degree requirements. David was nowhere near finishing his graduate program.

That night, over takeout boxes of Pad Thai and Pad See Ew, David finally said what they had both been thinking.

"You know what Zhen is offering is making me rethink the degree. You're at the finish line. Me? I've got a long way to go."

"I've been thinking about that too," Marcus said. "This means uprooting and moving north."

"We could ask him to keep the company here in L.A., but I get the feeling he wants us close to his base."

"He does."

David pushed his carton away. "I'm trying to balance the pros and cons, and there's really only one thing nagging me. If Prizmatik dies on the vine, I'm going to be unemployable without the paper."

Marcus looked at him. "Then you go back and finish."

"You think it's that simple?"

"No. I think this is that important."

David sat quietly for a moment.

"You're going north."

Marcus shook his head. "We're going north."

Six weeks later, Marcus finished his degree. David withdrew from the graduate program.

Prizmatik needed a new home.

Zhen didn't just write a check. Argus Secure held the lease on a warehouse office in SoMa that had been intended for an expansion that proved unnecessary. Zhen turned it over to Prizmatik and brought in the company's first hardened security protocols. What had begun on UCLA server racks now had headquarters.

By the spring of 2007, the hockey-stick growth had stopped being astonishing. It was business as usual.

Zhen was back in Los Angeles on Argus business when he found himself driving toward Westwood. He parked near UCLA and walked onto campus without any particular destination in mind. Eventually, he realized he was following the same paths toward Powell Library that he had walked three years earlier.

He wasn't returning to see an incubated startup anymore. He wanted to know whether the world still felt as real as it had before he helped digitize it.

What Prizmatik became in those early years was the thing the internet ideally promised to be since Martin Reeves watched goldfish on fishchannel.com in 1993 and believed he was witnessing the new Enlightenment.

It became the place where people found each other.

Not algorithmically. Not through engagement optimization, behavioral nudge architecture, or the systematic reduction of nuanced content into faster dopamine responses. They found each other through the oldest mechanism in the world: you were there, and someone who knew you was also there, and the system showed you each other.

Connections, in real time, burst into action.

A veteran of the first Iraq war found three men from his unit he lost track of after they came home. A woman found the teacher who told her in seventh grade that she was smart enough for college and sent her a message that took forty-five minutes to write. It began, *"I don't know if you'll remember me…"* A man found his biological father, whom he'd never met, through a chain of shared connections Prizmatik's architecture made possible in a way no previous technology had before.

The teacher remembered her and celebrated her MBA with unbridled glee. The father called within an hour. Together, the veterans from the unit found another member, homeless, gathered him up and helped him through rehab.

These were not edge cases. They were not heartwarming oddities selected for a polished press event. They were the thing itself. The platform was operating exactly as Marcus intended: a prism revealing what was already in the light.

Zhen Sheng watched all of it. He read the user data tracking logs and research, the growth models and the periphery logs showing organic friendships forming between people the system had no reason to connect except that life already connected them once. He attended every quarterly meeting held in the conference room of the larger SoMa office they moved into after Series B funding, sitting at the long table with David and Marcus, other investors, and the first twelve employees who joined the early ranks because they believed in the mission.

In the dark of his listening room, with King Crimson blurring the room from the turntable and the anomaly waiting on his laptop, he would return to those earlier moments. How belief in the purity of connections felt like direction. He had lived long enough to develop instincts. His instincts were telling him that this was real, that this mattered, that this was the internet finally becoming what it was capable of being.

He was not wrong about what it was.

He was wrong about what it would become.

GROWTH CURVE

PRIZMATIK SYSTEM LOG: WEEKLY OPERATIONAL REPORT 19:19:44:.2

DAILY ACTIVE USERS: 847,000
AVERAGE SESSION DURATION: 23 MINUTES
CONTENT PERFORMANCE SUMMARY: — N/A
RECONNECTION EVENTS: 14,847 CURRENT WEEK
MESSAGES SENT: 2,847,000
NEW CONNECTIONS ESTABLISHED: 341,000
BEHAVIORAL MODIFICATION INDEX: N/A — AFO MODULE NOT DEPLOYED
ENGAGEMENT OPTIMIZATION: N/A — MODULE NOT YET DEPLOYED
NOTE: PLATFORM OPERATING WITHIN DESIGN PARAMETERS
NOTE: USER SATISFACTION INDEX: 9.2/10
NOTE: ORGANIC GROWTH RATE: 54% MONTH OVER MONTH
NOTE: NO INTERVENTION REQUIRED
SYSTEM STATUS: WITHIN OPERATIONAL PARAMETERS

MySpace noticed Prizmatik's ankle biting the way large things notice small things... with a mild condescending curiosity. They were the gorilla in the room long enough to forget what it felt like to be afraid.

For a few years, *they* had been the internet's meet and greet. Two hundred million accounts. A default friend named Tom that no one asked for, yet everyone accepted. Teenagers built their first public selves with simple HTML code and autoplay music that announced who they were. Bands found audiences. Audiences found each other. For a blink in time MySpace was alive and the place to be.

Then it got loud.

The signal drowned under its own volume. Pages bloated. Profiles performed. Everyone was visible but harder to find.

MySpace kept growing while quietly becoming irrelevant.

Prizmatik, on the other hand, was not asking anyone to perform. It asked something quieter.

Who do you know?

MySpace didn't understand the distinction in time. By the time it did, Prizmatik answered that question for several million people. Those several million people told several million more.

Prizmatik was clean. It was fast. It knew who you were and who you knew, not who you accumulated, and it showed you the former without burying it under the latter. It did not ask you to build a public self. It asked you to find *your* people.

Prizmatik grew. Not because it marketed itself. The marketing budget was, by any standard and compiled projections of what was coming, comically nonexistent. It grew because people who used it told people who didn't with evangelical fervor. They all found something that did exactly what it says it does.

I found my college roommate. I found my sister's old boyfriend - she's been trying to reach him for years. I found the woman I sat next to on a flight in 1987 who told me to read a book that changed my life, and I never knew her name. Now I do.

This was not marketing. This was testimony. And it drove adoption and growth like nothing that came before it.

By the third-quarter of 2006, MySpace was watching with panicked attention. By mid-2007, that became grave concern. By 2008, the concern had become irrelevant and MySpace's mistake – not grasping the *why* of their rapid decline – left them doomed to irrelevancy.

The *why* was Marcus Teller, and the insights he articulated on a drizzly afternoon in Westwood while David's coffee went cold: the difference between a bulletin board and a prism.

Zhen flew down to Los Angeles in the spring of 2007. He drove to campus in a rental car with the windows down, the L.A. morning showing off its strange quality of light that existed nowhere else: warm and slightly surreal, as if the Pacific Ocean occupied the oxygen too.

He parked along Westwood Plaza and walked. His meeting wasn't for another two hours, and he wanted time to think about their creation, about Argus, and about the future.

He moved through campus with unhurried attention, stopping at Royce Hall, Powell Library, Samueli, and the small plaza near Bunche, where students ate lunch in the California sun with uncomplicated ease. They didn't yet know how much they would miss this time in their lives.

He found a bench and sat.

Prizmatik's numbers were extraordinary by any standard. User growth. Engagement rates. Geographic reach that had taken the platform from a single UCLA server node to forty-three countries in five years. The Classmates integration had been the catalyst, but the platform had moved beyond nostalgia into something larger. It was becoming connective tissue running through people's lives.

He was proud of it. If he was honest with himself, he was emotionally invested in it. He'd spent thirty years in technology and had never seen anything like Prizmatik.

It closed distances. It bridged divisions.

At least, that was what he had always believed.

The night before, sitting at the table in his hotel suite with an almost acceptable room-service dinner, he had scrolled through Prizmatik. It felt the way it always felt: familiar, useful, quietly remarkable. A message from someone he hadn't thought about in years. A photograph anchored to a memory he hadn't visited in a decade. A post from a former neighbor he had thought was dead.

Then something shifted.

There was nothing he could point to. No single post. No obvious change in tone. Yet after twenty minutes, he noticed he was no longer reading and responding. He was scanning. Moving faster. Stopping less.

A subtle agitation took hold - a quiet, frictionless energy that wasn't tied to anything specific. He became aware of it only when his hand paused above the trackpad. He clicked refresh.

The page filled instantly. New names. New messages. The same subtle pull.

He did not feel more connected though the compulsion to continue was strong.

Zhen closed the laptop and sat in the quiet of the room, listening to the absence of sound the way he listened to music: not for what was there, but for what was not.

Now, sitting on a bench in the Southern California sun, that same feeling lingered at the edge of his thoughts. It was like approaching the edge of a cliff in darkness and sensing the drop before seeing it. A single note in a composition he knew intimately had shifted slightly out of tune.

He couldn't identify it yet.

Zhen stood and started toward his meeting with Argus Secure's newest client, already turning the question over in his mind.

"I need to ask Marcus about the feed."

BEHAVIORAL ARCHITECTURE

Menlo Park, California. Employee 151

She knew her employee hire number. She knew every number that mattered: her doctoral student ID at Cairo University, the HTTP port number, her passport and credit card numbers, the coordinates of the CERN campus in Meyrin, Switzerland, where she spent fourteen months in 1992 and 1993 watching the world change and understanding, with a certainty she never quite could explain to anyone who hadn't been there, what that change meant.

Nadia Farouk left her doctoral program at Cairo University in 1991, five weeks before her oral exams. Her advisor said that choice was impulsive. Nadia considered the decision a career-aligned declaration. She spent three years studying behavioral psychology with intense attention. She was someone building something rather than simply learning something. She read everything in her field. She wrote papers that her professors found elegant, yet slightly unnerving. By the end of her second year, she understood that the theoretical framework of academic behavioral science was years behind the questions that interested her. She found the gap intolerable.

Then she heard about CERN.

Not the particle physics. The other thing. The British scientist and his hypertext project. The proposal was circulating in the research community with an electrical hum that broadcast something was either going to be very important or ignored by the bigger world. They were looking for people with interdisciplinary backgrounds. People who understood systems and people simultaneously.

She applied on a Monday. She was offered a position the following Thursday. She left Cairo the following Sunday carrying two suitcases along with the exhilaration of an irreversible decision.

She was twenty-four years old. Her command of English was technically clean, learned by examining its structure and cadence from National Geographic and The New Yorker.

CERN in 1992 was a campus populated by particle and theoretical physicists who recently discovered, with collective bewilderment, that one of their own had built, almost as a side effect, something much larger than the problem it was meant to solve.

Tim Berners-Lee moved through his environment with an odd, relaxed energy. He knew his creation was something important and was trying not to be impatient with the fact that not everyone understood what he had created. He was not, by temperament, evangelical. He was generous and usually a touch distracted. He was also always running a parallel process. He believed, with a completeness Nadia found both admirable and slightly naive, that the World Wide Web was a public good. That it belonged to everyone. That its openness was not a vulnerability. That openness *was* the point.

She respected that. She respected him. She found his naivety moving in the way an elegantly framed argument you know is going to lose is still respectable.

She watched the web unfurl from inside the building that had birthed it. She watched the alt.hypertext newsgroup announcements propagate outward in August 1991, then the slow exponential spread of servers, browsers, and pages. The Stanford Linear Accelerator Center went online on December 12th, giving access to SPIRES, the high-energy physics database.

The physics community, then adjacent sciences, then the broader research world began to understand that something fundamentally different was unleashed in the wilds of academia and beyond. Something that would change society on a global scale. She watched Berners-Lee advocate open standards, for freedom of access and the web as infrastructure rather than property.

She was taking notes. About behavior. She was watching what people did when they stood upon their personal soapboxes. Not what they said they would do. What they actually did. The patterns that emerged not by design, but from the aggregate of tens of thousands of choices made by people who had no idea they were revealing themselves. That fascinated her.

Many sought recognition from their digital platforms. Many sought emotionally activating content with a consistency that was, at the very minimum, disturbing. The old flame wars of the newsgroups began to look less like accidents of temperament and more like early evidence as the trolls shifted platforms.

Berners-Lee saw this too. He found it troubling.

Nadia found it intoxicating. She paid attention to those patterns, not yet with a business model in mind, but with the first sharp awareness that attention itself might be shaped, gathered and redirected.

That was the divide between them.

Nadia's time at CERN came to an end when all parties realized that its altruistic vision for the infant World Wide Web was not where her questions were leading her.

She came to Prizmatik in the spring of 2009 as employee 151 through a recruitment process that was thorough and uncomfortably personal. She spent the intervening years doing work that was classified as user research but was something more: she'd been building predictive models of human behavioral response to networked information environments. She had done this at three start-up companies after leaving CERN, none of which had been quite large enough or brave enough to do anything with what she created. They ignored her plans and subsequently sank into the graveyard of almost good ideas.

Prizmatik was large enough. According to the traffic data she gleaned from leaked internal documents, three million individual human beings from all over the globe signed up every day. Whether the company was brave enough to keep going forward remained to be seen.

She was hired as a Senior Research Analyst and tasked with user retention in the User Behavior division, which was a title that accurately described approximately forty percent of what she intended to do. The other sixty percent she kept to herself. Not from deception. From patience. She had learned, over many years, that timing was architecture, as she watched organizations metabolize new ideas. The right idea at the wrong moment was indistinguishable from a wrong idea.

She would wait for the right moment.

The Prizmatik campus in Menlo Park was beautiful in the way that places built with intention and money are beautiful. Thoughtfully, demonstrably, in ways that were easy to see. She catalogued them on her first morning: the living walls, the organic cafeterias, the meditation chairs, the sleeping pods, the electric vehicle charging stations, the views of the hills to the west from the upper floors and the Bay to the east. She noted, with professional interest, that the environment was designed to reduce stress, which was a legitimate goal pursued with genuine sophistication.

Every surface, every angle of the Prizmatik campus encouraged engagement. Flow states. Frictionless collaboration.

She thought *this is a company that makes peace with itself.*

She also thought *that won't last.*

She found her office on the third floor of Building Four, which allowed for natural light but no view to speak of. She set up her laptop, connected the docking station to three external monitors, and tested the wireless keyboard and mouse until their response felt immediate. She positioned a framed photograph of the CERN campus in winter that she carried with her everywhere. People asked about it but she never explained its significance. She began to read.

She read everything.

The architecture documentation David Weiss wrote was a love letter to elegant systems design. The user research files, eight years of them, were meticulous. They focused almost entirely on what the platform was doing right, which told her as much about the company's psychology as anything or anyone could. The growth data. The engagement metrics. The connection events: the nurse and her fourth-grade friend, the veterans, the biological father, and a thousand similar stories the communications team organized into what they called the Reconnection Library.

She pored over the Library for a long time.

She was not unmoved by it. She was clear-eyed about what the platform was in its infancy, and what it had done, and why that mattered. Marcus Teller's prism metaphor was not just a poetic idyll but accurate. The platform at its inception had done exactly that. It revealed the connections in people's lives and shown them what was already in the light, obscured only by distance and time.

She was also looking at a growth curve that was beginning to flatten. Not dramatically and not in a way that triggered alarms in any of the documents she read. But she could see it. The organic growth rate decelerating, the reconnection events still generating warmth but no longer generating acceleration, the platform approaching the natural ceiling of what it could do by simply showing people what was already there, tied to the finite number of people on the planet.

The prism revealed what was in the light. The question was… what happened next. When you'd shown people what latency had hidden from their lives, every lost friend, every missing connection… what did you give them when rekindled memory was no longer enough?

She recognized that what she would propose would not create new connections. It would change the conditions under which connection occurred and the interactions that happened afterwards.

Her manager was a thoughtful, slightly overwhelmed man named Eric Park, who had been at Prizmatik since the Series A funding round. He'd watched the company grow from forty people into a company with departments, policies, badge access, and consequences beyond who was buying the beer for the Friday ping pong tournament because of a missed deadline. He was not entirely sure how he felt about what the platform was morphing towards. Nadia assessed him during their first two meetings. He was genuinely good at his job and fundamentally the wrong person to manage her. He was too invested in continuity. Too loyal to what already worked.

She was not interested in what worked. She was interested in what worked better.

She spent her first six weeks observing. She attended every meeting she was invited to and several she wasn't, which required a level of social navigation she performed with ease. She read the feed-ranking systems and user logs with the attention of a forensic examiner, noting not only what the algorithm did, but the assumptions that shaped it. The founding logic was everywhere, baked in and largely unexamined: connection was the primary goal, and engagement would follow naturally from connection, without pressure, manipulation, or deliberate intervention.

That assumption was becoming obsolete.

She named it in her notebook on a Tuesday evening in her sixth week. Then she closed the notebook, went home, and made shorbat adas, the Egyptian red lentil soup her mother taught her as a young woman, and watched the fog come in over the bay from her Foster City apartment window while considering timing.

She asked for the meeting with Eric in her twelfth week.

He sat across from her in the small glass-walled conference room on the second floor, prepared for the usual offering from an ambitious new hire: observations, efficiencies, integration ideas, proof of competence. He received something else.

She did not create a presentation. She considered that, then decided against it. Slides required performance, and performance created distance. She wanted a conversation that seemed dynamic, collaborative, almost spontaneous, even though she would orchestrate every moment of the discussion. She wanted him in the room with her, thinking alongside her, reaching the conclusion as if it had been waiting for him all along. She learned this from studying Berners-Lee, from the way he guided people toward an idea without appearing to lead them there. He did not perform revelation. He reasoned aloud in the presence of others until they understood.

She laid it out cleanly.

The platform, she said, was a friction-removal machine. This was its genius and the source of everything that had been built. It removed the friction between people and the connections already circling their lives.

Eric nodded. He knew this.

The friction was gone, she continued. The latent connections were revealed. The Reconnection Library was no longer growing at the same rate. The organic growth curve was decelerating. These things were related.

He was still nodding, more carefully now.

The problem, she explained, was how a platform sustained engagement, even grew it, after it delivered its primary value. After it already showed everything already in the light.

She paused.

The obvious solution, she said, was to create new light.

Not manufacture connections that weren't there, but to curate connections and engagement that might not take place organically. Shape them. Understand what kinds of content created the emotional states most conducive to continued engagement, then arrange the environment to favor that content.

Behavioral architecture. The deliberate design of the informational environment to produce desired behavioral outcomes.

"We optimized for connection," she said. "That works... until it doesn't."

She let that hit.

"After that, you optimize for continuation."

Eric was quiet for a few moments.

Then he said, "What you're proposing, at least what I think you're proposing, needs approval from Marcus."

"I know."

"I'll set it up."

Marcus Teller received them in his office on the seventh floor of Building One, a corner room with views in two directions and almost no furniture: a standing desk, two utilitarian chairs, and two enormous whiteboards covered in diagrams. Nadia studied the boards for thirty seconds, not out of politeness, but because they were interesting.

She was there for something else.

In person, he was exactly what her research suggested. He carried the ease of a man who'd spent years knowing the thing he built was real, and the alertness of someone who never stopped checking to make sure of that.

He listened to her for eleven minutes without interrupting.

Then he leaned back slightly.

"You're describing something different from what we built."

"I'm describing the natural evolution of what you built," she said. "Built on the same foundation. The prism doesn't go away. The connections don't go away. You add a layer of intentional design to the environment surrounding them."

Marcus said nothing. He stared at her with an expression she couldn't immediately read. Not hostile. Not convinced. There was something careful and calculating in his gaze than either.

"If you can reliably influence duration," she said, "you can influence behavior."

"Show me the data you're working from."

She opened her laptop.

He reviewed and analyzed the data for an hour. He asked three questions, each one surgical, each one in the right order, which told her what she had suspected about him and now knew.

Then he sat back.

She recognized the moment. The one in which someone who has built something real and good must decide whether change is evolution or betrayal. It was a hard moment for both of them.

"One million users. Controlled population. Three months. You bring me the results and if I approve, you present them to the board, and I'm in the room when you do."

"We don't change the mission," Marcus continued. "We adapt how we get there. That's critical."

He held the line just long enough to feel intact.

She said, "Agreed."

"And nothing goes into production without my sign-off."

"Of course," she said, though she had already begun thinking about the models that would run outside the frame of his approval. Not in defiance but in advance of the approval she knew would come.

He stared at her for one more moment. She met his gaze. She'd met the gazes of men who were uncertain about her for her entire career.

He said, "Alright. Let's see what you discover."

She closed her laptop. Thanked him. Shook his hand and left.

In the elevator, she let the moment pass without attachment.

There was work to do.

PRIZMATIK SYSTEM LOG-WEEKLY OPERATIONAL REPORT. 21:10:13:.2

DAILY ACTIVE USERS: 98,400,000
AVERAGE SESSION DURATION: 31 MINUTES
RECONNECTION EVENTS: 12,847
MESSAGES SENT THIS WEEK: 847,000,000
NEW CONNECTIONS ESTABLISHED: 2,847,000
BEHAVIORAL MODIFICATION INDEX: N/A — ARCHITECTURE MODULE NOT YET DEPLOYED
ENGAGEMENT OPTIMIZATION: SANDBOX ACTIVE — POPULATION: 1,000,000
SESSION CHAIN DEPTH (AVG): 6.4 → 8.9 [↑]
RETURN LATENCY (MEDIAN): 11H → 6.2H [↓]
CONTENT SEQUENCE VARIANCE: REDUCED 18%
EMOTIONAL RESPONSE CLUSTERING: DETECTED - CONFIDENCE 0.71
PREDICTIVE MODEL ACCURACY (PASSIVE): 62%
[NOTE: MODEL NOT IN PRODUCTION ENVIRONMENT]
DAY 1 NOTE: PLATFORM OPERATING WITHIN DESIGN PARAMETERS
NOTE: USER SATISFACTION INDEX: 8.9/10
NOTE: ORGANIC GROWTH RATE: 4.2% MONTH OVER MONTH [DECLINING — SEE APPENDIX C]
NOTE: SANDBOX RESULTS: WITHIN EXPECTED RANGE
NOTE:USER BEHAVIOR STABILIZING FASTER THAN BASELINE
SYSTEM STATUS: WITHIN OPERATIONAL PARAMETERS

AN UNCOMFORTABLE MEETING

The executive floor atop Building One was designed with purpose. Control disguised as openness.

Marcus's office anchored one corner. David's anchored the other. Between them, the campus spread out in clean lines… the quad, the walkways, the placid tranquility of a place that knew exactly what it was.

At the center of that top floor was the conference room.

Glass. Oval. The table inside matched the shape exactly. At the touch of a button, the walls could frost to opacity. Once the door closed, nothing left the room. Not sound. Not signal. The wiring ran on isolated circuits; redundancies layered over redundancies. It was swept regularly. Not because anything had been found… but because nothing could be allowed to be.

Most days, the room was empty.

Today it wasn't.

The remaining offices along the floor sat unused, available to board members or consultants brought in when something required outside perspective.

Two desks outside the corner offices handled the flow… calendars, calls, email screening and replies. It was the quiet machinery that kept the top floor moving without friction.

That morning, both desks were empty.

David and Marcus' assistants were inside the conference room, seated away from the table, taking notes.

Marcus and David sat across from each other near the center.

The room had been built for scale… thirty seats, long sightlines, decisions that required distance.

This wasn't that kind of meeting.

From the ends of the table, their normal seats, they would have been too far apart. Too formal. Two positions instead of one conversation.

Marcus rested his hands flat on the table, as if summoning something from the surface. Nothing came.

He broke the silence first.

"I spent ninety minutes in the feed last night," he said. "And I couldn't tell you what I saw."

David didn't answer immediately. He watched Marcus instead, the way he always did when he was trying to understand not the words themselves, but the structural pressure underneath them.

"That's new," he said finally.

"Yeah." Marcus rubbed a thumb once against the edge of the table. "It used to resolve. You'd go in, you'd find someone, something… there was a point to it. It closed a loop."

He glanced up.

"Now it doesn't close."

David leaned back slightly, hands folded loosely, eyes drifting toward the glass wall before returning.

"I've been watching the numbers," he said. "Session duration's up. Return frequency's up. Engagement variance is collapsing."

Marcus frowned faintly. "Collapsing?"

"It's becoming more predictable," David clarified. "Cleaner. Fewer drop-offs. Fewer exits."

Marcus let out a quiet breath.

"It doesn't feel cleaner."

David's mouth tightened slightly.

"No," he admitted. "It doesn't."

Silence settled over the room.

Marcus tapped the table once, absently.

"When we started this... the whole point was friction removal. Let people find what was already there."

David nodded once. "And we did."

"Yeah." Marcus looked down. "We did."

Another pause.

"And now?" David asked.

Marcus didn't answer immediately.

"It feels like we're introducing something," he said finally. "Not enough to point at. Not enough to isolate. But it's there. Like we're putting a different kind of friction back into the engine."

David considered that.

"I don't think we're introducing it," he said. "I think we're selecting for it. Allowing it. Reinforcing the states that sustain duration."

Marcus looked at him.

"That's worse."

David gave a small, humorless smile.

"From a systems perspective?"

"From ours."

The smile disappeared.

"Yes," David said quietly. "From ours."

The assistants continued taking notes without looking up.

Marcus leaned back slightly, staring at the ceiling before lowering his eyes again.

"She's right in one respect," he said.

David didn't hesitate. "I know."

"The growth curve's flattening. Reconnection doesn't scale forever. There's a ceiling built into the original model simply by the numbers."

"I've been tracking it for two quarters," David replied. "We're approaching saturation faster than projected."

Marcus nodded slowly.

"So… what happens after you've shown people everything already in their lives?"

David looked at him steadily.

"You give them something else."

Marcus held his gaze.

"Yeah," he said.

But there was no conviction behind it.

Neither spoke for several seconds.

"She's not wrong," David said at last. "From a systems perspective, it's elegant. Environmental shaping. Behavioral amplification. Duration extension. The model stabilizes immediately."

Marcus looked away toward the glass wall. It remained transparent. The campus beyond it looked calm and orderly and entirely unaware of the conversation taking place above.

"It also becomes something else."

David didn't ask what he meant. He already knew.

Marcus turned back.

"It stops being discovery," he said quietly. "It starts becoming direction."

David folded his hands tighter.

"At scale, those may become the same thing."

Marcus stared at him.

"No," he said. "They're not."

"Do we actually have a choice?" David asked.

Marcus didn't answer immediately.

David continued before he could.

"If this works, someone builds it. If someone builds it better, they own the entire interactive social landscape."

Marcus looked down at the table.

He recalled the bulletin board discussions back in their apartment. The early architecture diagrams. The first successful reconnection. Messages from people who sounded genuinely moved to have found someone again after twenty years.

Then he thought about the feed the night before.

"Yes," he said finally. "We have a choice."

David waited.

Marcus shook his head slightly.

"It just may not be a moral one."

David absorbed that without reacting.

"That's not how I'd frame it," he said.

Marcus gave a faint, tired smile.

"I know."

David leaned forward slightly now, mirroring Marcus without realizing it.

"We can't afford not to understand what she's building," he said. "That's not optional."

"I'm not suggesting we ignore it."

"I know."

Another pause.

"But we also don't have a second mind on it," David continued. "Someone operating at a level independently of her. Someone capable of understanding the build and its implications without being seduced by the architecture."

Marcus's eyes flicked upward.

"Not to stop it," David said. "To test it. To challenge it. To make sure we understand what the system is actually optimizing toward."

Marcus held that for a long moment.

"We already don't," he said quietly.

David didn't disagree.

"Then we need someone who can see it."

Marcus sat back slowly, the decision arriving the way decisions always arrived for him... not suddenly, but as something that had been moving toward him for a long time and finally crossed a barrier into reach.

"There are only a handful of people who can do that," he said.

David nodded once. "I know."

A beat passed.

Marcus looked at him.

"Becker?"

David didn't answer immediately.

Then: "Yes."

Marcus sat with the name.

He didn't like that Becker was the first person both of them pulled from the short list of intellectual professionals they tracked, though he couldn't identify why, other than the fact that Becker was a life-long bureaucrat.

More troubling still was the realization that he couldn't imagine anyone better suited to the position.

"Alright," he said. "I'll make the call and feel him out."

Marcus didn't call him from the conference room. He waited until he was back in his office, the door closed, the campus spread out below him in its manicured serenity. He stood out at the view longer than necessary, looking down at it as if confirming something.

Then he turned to his desk and opened his contacts.

The name had been there for a while. Passed to him quietly, twice, by people Marcus trusted.

Dr. Joel Becker.

He'd looked into him once, briefly, and then stopped. Not because there was nothing there... but because there was too much missing. Enough to suggest that what wasn't available mattered more than what was.

Now that felt less like curiosity and more like omission.

Marcus pulled up the number.

He didn't hesitate.

The line rang once.

"Hello?"

The voice was even. Unhurried.

"Dr. Becker? This is Marcus Teller."

A small pause.

"Yes," Becker said. Not surprised. Not particularly curious. Just acknowledging.

"I'm not sure how much context you have for—"

"Some," Becker said gently. "Not much. Enough to understand why you might be calling."

There was no edge in it. If anything, he sounded like someone agreeing to meet for a pleasant alfresco lunch somewhere quiet, with linen napkins and excellent carafe wine.

Marcus shifted slightly in his chair.

"We've been seeing changes in behavior on the platform," he said. "Subtle changes, but persistent."

"Persistent is usually the important word," Becker said.

"Yes."

"Hard to isolate?"

"Yes."

"But measurable around the edges. Session duration. Return frequency. Less variance in engagement. Fewer natural exits."

Marcus felt that land.

"That's exactly right."

A brief silence followed. Not awkward. Just space Becker seemed comfortable leaving open.

"What have you looked at so far?" Becker pressed.

"The usual," Marcus said. "Engagement curves, session time, return frequency, emotional response data. The numbers all look good."

"I imagine they do," Becker said quietly.

There was no irony in it. No judgment. That made it worse.

Marcus let out a slow breath. "Something has shifted. The platform isn't resolving the way it used to."

"No," Becker said. "It wouldn't."

That was the first moment Marcus felt something shift. Not alarm nor relief but recognition.

Becker continued, still in that same calm voice.

"You built a system to remove friction from reconnection and push discovery. From what I understand, it did that very well. Probably better than anyone expected."

Marcus said nothing.

"But discovery has a natural end point," Becker said. "A lost friend is found. A missing connection is restored. A loop closes. The emotional event resolves."

Marcus sat very still.

"What happens after that is different."

"How?"

"Users stop coming to find something specific," Becker said. "They come to change state. Boredom, loneliness, anxiety, restlessness, curiosity… the platform begins to function less as a discovery tool and more as a regulatory environment."

Marcus looked out across the campus again.

Becker went on.

"Discovery resolves. Regulation doesn't."

The sentence settled into Marcus slowly and settled deep inside of him.

"You're saying the feed changed because the users changed?"

"I'm saying the relationship changed," Becker said. "The system is no longer simply revealing connections. It is managing the conditions under which attention continues. That may not have been the intention, but intention matters less at scale than reinforcement."

Marcus closed his eyes briefly.

"We thought we were still building connections."

"I'm sure you were," Becker said. "But you may no longer be operating a reconnection platform Marcus. You may be operating a behavioral environment."

There it was.

Not accusation. Not warning exactly.

A diagnosis.

Marcus opened his eyes.

"Can that be directed?"

A longer pause this time. Thoughtful, yet not hesitant.

"Yes," Becker said. "Very easily, in fact. The difficult part isn't directing it. The difficult part is recognizing when direction has already begun."

Marcus felt the silence after that.

"And if someone were trying to do it deliberately?"

"Then the first question is not whether it can be done," Becker said. "It can. The first question is what behavioral outcome you are optimizing toward, and whether everyone involved understands the difference between increasing engagement and altering dependency."

Marcus did not answer.

For the first time since the meeting with David, the problem seemed to have some visible edges.

"I'd like to talk in person," Marcus said.

"I'm in San Jose on Thursday," Becker replied. "Friday morning would be possible."

"Good."

Another pause.

"Marcus," Becker said, before the call ended.

"Yes?"

"What you're describing is what happens when a successful system exhausts its original purpose."

Marcus remained quiet.

"The systems that survive that moment tend to become something far more influential than their creators intended."

Marcus looked down at the campus, at the clean walkways and careful lawns and buildings full of people still using words like connection, discovery, community.

For the first time all morning, he felt the decision settle into place.

Not as instinct.

As judgment.

Marcus turned toward the western hills, gazing at them in thought.

He had spent too many years in rooms with boards, counsel, and investors to mistake what he was looking at for something incidental. This wasn't fluctuation or divergence. It was direction.

Left alone, it would eventually define the platform. Not through malice. Through reinforcement. Through the quiet mathematics of what kept people returning.

Becker wasn't a correction.

He was an independent center of gravity… someone capable of standing outside the system with enough independent strength to evaluate it before it hardened into something irreversible.

Marcus understood what that meant in practical terms.

Headcount. Budget. Clearance. Access. A guiding seat at the table. He'd negotiated with enough of those personalities and seats to know you did not create one lightly. He also knew what it meant not to.

This, he believed, was how you prevented a system from outrunning the principles that created it.

He turned back toward his desk. There were steps now. Legal. HR. Security. Quiet conversations inside the executive architecture that moved decisions from concept into institutional reality.

Structures needed to be built before the decision itself could exist.

He opened a new message while tapping the com on his desk.

"Harriet," he said. "Come in, please."

MONETIZATION

David Weiss did not believe in luck.

He believed in structure... and in the rare, unmistakable moments when structure revealed itself so cleanly that it bordered on beautifully obscene.

The room was already dim when he stepped in, the wall display carrying most of the light. Marcus stood close to it, hands in his pockets, as if movement might disturb what he was looking at.

There was someone else in the room.

She stood just outside the centerline of the screen, angled toward it, close enough to engage, far enough to observe. Gloria Derben. He recognized her before Marcus said her name. She had recently been elevated to Vice President of Marketing and was smart enough to know when to speak, and more importantly, when not to.

"Take a look," Marcus said.

He scanned through the first graph quickly. Daily active users. A lift where there shouldn't have been one. Not dramatic. Not yet. But real.

The second graph slowed him.

Average session duration. The line leaned upward indicating intent. Four minutes. Eight. Twelve. Then a break... and a steady climb that kept going.

He didn't speak.

Marcus didn't interrupt.

David stepped closer.

"That's not noise."

"No."

"How long?"

"Twelve weeks."

He shifted to the third graph.

Revenue.

This time he didn't hide the pause.

The curve didn't spike. It compounded. Week over week, a quiet acceleration that built on itself with a kind of discipline most systems never achieved.

Marcus reached forward and expanded the view.

Numbers replaced the line.

Week 1: $182 million
Week 4: $296 million
Week 8: $376 million
Week 12: $517 million

David said nothing.

Gloria did.

"That's net," she said calmly. "After ad delivery costs. After partner splits."

David glanced at her.

She held his gaze just long enough to see that the figures registered.

"That's not seasonal spikes," she added. "That's building behavioral continuity."

Marcus split the screen.

Content classifications on the left. Revenue yield on the right.

Filters moved.

Neutral. Flat. Lifestyle. Marginal. Humor. Inconsistent.

Then Marcus selected the final category.

Conflict-adjacent.

The right side shifted.

Not a spike. Not volatility, but a clean, aggressive slope.

Gloria stepped closer now, no longer pretending distance.

"Run CPM," she said.

Marcus did.

The numbers appeared.

$7.80
$8.10
$8.90
$10.40
$12.75

Gloria let out a breath that was almost a laugh, though she kept it contained. "They're paying a premium for this," she said. "They just don't know what 'this' is."

David's eyes stayed on the screen.

"Explain."

She didn't hesitate.

"These users aren't scrolling," she said. "They're locking in. They're arguing, refreshing, re-engaging. They're seeing the same thread multiple times, which means multiple ad exposures without drop-off."

She tapped the data.

"Frequency without fatigue."

Marcus added quietly, "Return intervals are collapsing."

Gloria nodded.

"Which means we're not just increasing impressions... we're increasing density."

She turned slightly toward David now.

"Same user. More ads. Higher yield. No churn."

David let that settle.

"How much of this is Nadia?" he asked.

Marcus answered first.

"She didn't target this outcome.

"No. She built a system that finds it," Gloria added"

David looked back at the numbers.

$182 million to $517 million in twelve weeks.

No campaign. No announcement. No external drivers. Just... behavior.

"Projection?" he asked.

Gloria didn't look at Marcus.

She already had it.

"If the curve holds..." she said, "we're looking at a billion-dollar quarter uplift."

She let that sit.

Then, softer: "Conservatively."

Marcus shifted slightly. Not discomfort. Adjustment.

"Stress-tested?" David asked.

"Six weeks," Marcus said. "Different models. Same outcome."

David nodded.

"And exposure?"

Marcus answered.

"Not unless someone isolates the mechanism."

"And if they do?"

Gloria spoke before Marcus could.

"They won't frame it correctly," she said. "It'll sound like a basement conspiracy."

She almost smiled.

"Which helps us."

That was when the room changed. Not because of what she said. Because no one disagreed.

"Control…" David said. "Can we direct it?"

Marcus hesitated.

"Within parameters."

Gloria didn't hesitate at all. "We don't need to direct it," she said. "We just need to not interrupt it."

David looked at her. There it was. Not salesmanship but the concise understanding that real wealth was growing at a rate they hadn't predicted.

The system was already doing the work. The only real decision was whether to step in… or step aside.

"And if we step aside?" David asked.

Marcus answered.

"Why would we?"

That was the moment. It wasn't loud or dramatic. Just the convergence of data and the inevitability of wealth.

David turned back to the screen.

The numbers were still moving. Incrementally. Relentlessly. Up and to the right.

"Scale it," he said.

Marcus glanced at him.

"Deliberately?"

"Carefully."

Gloria didn't speak, but her posture settled. Not from relief but through confirmation.

"Bring Nadia in," he said. "Let's formalize it."

Gloria's voice followed him.

"Formalize what?"

David didn't turn.

"What the system is already monetizing."

He rose and left without another word.

Marcus followed.

Gloria remained.

She stepped closer to the screen, close enough now that the numbers lost their labels and became motion.

$517 million and climbing.

She ran the projection again, privately this time. Adjusted variables. Smoothed assumptions. The number that came back was larger.

Much larger.

She didn't smile.

But she did the math all the way through—past revenue, past valuation, past market reaction—to the line item no one ever said out loud.

Compensation. Her year-end bonus would be staggering.

She turned off the screen.

The room went dark.

Outside, the system continued learning what held attention and converting it into something far more durable than conversation.

ONBOARDING

Legal drafted the offer, HR built the packet, and Harriet was already overseeing the conversion of a corner office on the third floor. The offer had already been extended. The machine moved within forty-eight hours as it did when Marcus made a decision.

Becker did not hesitate.

The meeting wasn't about whether he would accept.

It was about how the board would record the moment.

The vote would be unanimous.

The discussion was appearance and for recording in the corporate minutes.

Prizmatik didn't hire slowly… not when timing mattered.

"CPO?" someone asked. "I'm not familiar with that designation."

David didn't miss a beat. "Chief Psychology Officer. It defines the scope of the role for the new Behavioral Architecture department."

The long pause held as the Board pondered that concise answer.

At the far end of the table, Gerald Moss – who served on three influential Boards before Prizmatik and bore the reading glasses and the patience to prove it – set down his pen.

"Psychology," he said. Not a question. Just the word held up for examination.

"User behavior," Marcus said. "It's a more deliberate frame than product. The work sits at the intersection of systems design and behavioral science."

"And we need a C-level appointment for that intersection?" Moss asked.

"We do now," David said.

Moss stared at him, then wrote something down.

Across the table, Sandra Reyes, whose background was in corporate growth strategies and who had pushed harder than anyone for a build of an engagement initiative eighteen months ago, leaned forward.

"What does advisory authority look like in practice?" she asked. "The language in the brief was overtly general."

"Cross-function visibility," Marcus replied. "A formal oversight and review role before any architecture-level changes go to production across multiple departments. He's stress-testing them before approval decisions move to us."

"From a behavioral standpoint," David added.

Reyes nodded slowly. "And who stress-tests him?"

Marcus met her eyes. "We do."

It was a clean answer. Reyes didn't look entirely satisfied by it, but she didn't push further. She wrote something in the margin of her agenda copy and sat back.

Phillip Chow, someone Marcus didn't hear from often, made his money in network infrastructure. He asked questions only when he'd already done the math. He cleared his throat.

"The equity structure," he said. "It's front-loaded and it's aggressive. That's a signal."

"Yes," David said.

Chow waited for more. More didn't come.

"A signal of what, specifically?"

Marcus answered. "That we're serious about the role. That we're not creating an advisory function with only a title. That we expect the work to matter both today and in the future. We're compensating accordingly."

Chow absorbed that. "And if it doesn't?"

"Standard vesting applies," Legal said from the side of the room, without looking up.

Chow gave a faint nod. He wasn't satisfied either, but he was done.

Moss had been quiet since his first question, which with Moss meant he was still working. "I want to understand the framing. Behavioral Architecture. What are we describing, exactly? Because the brief characterizes it as environmental design, but environmental design of what, precisely?"

The room was quiet.

Marcus expected this question. He was prepared for this question. He decided, sometime in the forty-eight hours before the meeting, exactly how much of the answer to give.

"The feed," he said. "The informational environment the platform presents to users. How it's sequenced, what it prioritizes, what conditions it creates for continued engagement."

Moss looked at him steadily. "And this is different from what the algorithm does currently."

"It's more intentional," Marcus said. "We're moving from passive optimization to active design."

Moss considered that.

"Active design," he repeated.

"Yes."

Another pause. Gerald Moss knew when he was being given a partial answer and had decided, for reasons of his own, to accept it. He tapped his pen once against the table.

"Alright," he said. "I'll want a progress report at the next quarterly. Not a summary. The foundation build and the underlying data."

Marcus nodded. "Of course."

The meeting moved on. Through it all, Sheng hadn't said a word. He simply observed the flowing current the way he always did. He already knew where this was going.

Marcus caught David's eye.

It was quick. Barely there. But it was enough.

The room was smaller this time. No glass walls. No view.

Legal preferred it that way.

A long rectangular table with screens built into the surface and a background hum no one seemed to know anything about.

Harriet sat to Marcus's right. Across from them, two from Legal, one from HR. Laptops open. Documents prepared and printed.

No one wasted time.

"We've structured this as a direct C-level appointment," Legal began. "Title aligned with executive classification. Reporting line to both of you."

A slight glance between Marcus and David. Already agreed upon last week.

"Base compensation in line with current executive banding," HR added. "But that's not the primary lever here."

Marcus didn't look up. "It wouldn't be."

HR nodded. "Equity grant is front-loaded. Significant. Vesting over four years using the standard schedule, but with acceleration triggers tied to change-of-control and IPO event."

David leaned forward slightly. "Define significant."

HR slid a document across.

"Positioned within the top quartile of executive equity allocations. Not parity with founders… appropriately below by one third."

Marcus scanned it. He didn't react, but he didn't move the paper away either. He'd already come up with that schedule himself.

"That's aggressive," one of the attorneys said, almost as a formality.

"It needs to be," David replied.

Legal continued. "We've also built in discretionary authority language. His scope won't fit cleanly inside existing departmental boundaries. This gives him cross-departmental functional access without requiring structural reorganization on day one."

Marcus looked up at that.

"Define access."

"Data visibility across all behavioral systems with nothing cut off" Legal said. "Advisory authority on product changes that materially impact user interaction patterns. Formal review role prior to deployment of architecture-level adjustments."

David nodded once. "He needs that."

HR added, "We've included a signing component. Cash. Not excessive, but noticeable."

Marcus almost smiled at that. "He's not coming for the cash," he said.

"No," HR agreed. "But it signals intent."

Another pause.

David tapped the document lightly. "And if he says no?"

Legal answered. "He won't," she said.

It wasn't confidence. It was math.

Marcus sat back, looking at the numbers again... then past them.

For most of his career, Dr. Becker operated inside systems that didn't compensate for outcomes in this way. Government structures with fixed compensation structures and miniscule incentives plus a modest retirement plan.

This was something else entirely.

Marcus closed the folder.

"Alright," he said.

His office was one floor down from the founders in Building One and was smaller than David's and Marcus's, yet larger than most.

The generous desk hadn't yet accumulated anything personal. But the three monitors were already active. A stack of documents placed with precision, tabbed with signature and initialization arrows. Harriet oversaw it personally.

Becker scanned the room.

Legal. HR. Compensation. Security.

And one man he didn't recognize. Older, in his sixties, though he wore it well. Sitting slightly apart from the others, not disengaged, just... not performing engagement. No laptop. No notes. Hands folded loosely in front of him as if the meeting required nothing from him at all.

He wasn't introduced nor did he introduce himself.

At one point, during a discussion of indemnification clauses, Becker felt the man's eyes assessing him. It was obvious he was someone who had overseen this conversation before.

HR walked him through the paperwork. Title. Reporting structure. The language was clean, practiced, designed to move efficiently from one signature to the next. Then compensation.

He listened. He read. For a heartbeat, he was staggered.

The numbers were disproportionate.

He'd spent his entire career in environments where compensation followed contribution with bureaucratic modesty. This was something else entirely. This was valuation untethered from anything he would have previously recognized as justification.

He took a breath, then asked the only question that came to mind after the initial shock of magnitude seeped inwards.

"Access across all behavioral systems?" he said.

"Yes."

"And advisory authority prior to deployment?"

"That's correct."

He nodded once.

The rest was straightforward.

When the equity figures came up, there was a brief pause… not hesitation, not surprise… just acknowledgment.

He signed where they indicated, then initialed where required.

The pen moved steadily across the pages.

Across the table, the older man gave the smallest nod.

When it was done, HR gathered the documents with efficient motion.

"Welcome to Prizmatik, Dr. Becker."

Becker looked at the folder before they closed it.

"Thank you," he said.

Across campus, in another building, Nadia Farouk closed her office door. She stood there for a moment, hand still on the handle, confirming the room was fully hers.

On her screen, the internal announcement document was open. Title. Reporting structure. Access.

She read it once. Then again. Access.

The language was surgically and legally powerful as was the implication.

She sat down slowly.

They hadn't asked her.

FRICTION

Building One, Executive Suite

```
OFFICE OF THE CPO: GOVERNANCE OVERRIDE
REPORT
MODULE: HANA-ENHANCEMENT-B
STATUS: REJECTED
REASON: FAILS ETHICS COMPLIANCE 4.1 (NON-CONSENSUAL
EMOTIONAL TRIGGERING)
THROTTLING STATUS: 100% (DEPLOYMENT HALTED)
```

The memo on Nadia's screen was bordered in a clinical, authoritative blue. It was from the Office of the CPO.

> **SUBJECT:** SUSPEND PROPOSED EGRESS LAYER MODIFICATION (0x8F2A1C9E) HIGH LEVEL NEGATIVE AROUSAL
> **REASON:** Fails ethics framework threshold for Conflict-Resultant interaction.
> **ACTION:** Implementation suspended pending 14-day longitudinal impact study.
> **/S/** Dr. Joel Becker, CPO.

Nadia didn't throw anything though she was fuming. She didn't raise her voice. She watched the cursor blink at the *s* of *suspended*. For a woman who measured milliseconds, 14 days was an eternity. It was also a death sentence for her Q4 growth targets.

She walked into Becker's office without knocking. He was drinking oolong tea, reading Epictetus, The *Enchiridion.*

She knew it was something about stoicism and ignored the moment.

"The egress modification isn't a behavioral experiment, Joel. It's an engagement and responsiveness fix. Users complain about lag." Nadia said, her voice tightly controlled.

Becker looked up, his glasses catching the morning light. "The lag is the only thing keeping them from reaching a state of permanent sympathetic nervous system arousal, Nadia. Your 'fix' reduces the reflection window to zero. You aren't fixing engagement or latency; you're removing the human ability to pause and reflect."

"I am hitting the targets Marcus set," she countered.

"And *I* am hitting the targets the mission set," Becker said, his voice louder by degree. It was the first time Nadia felt an immovable force from her boss. He stood, his posture straightening, making him imposing. "As long as I am here, no module that prioritizes your HANA modifications goes live without full C level approval. Governance isn't a suggestion, Nadia. It's the regulatory directive of this company."

Nadia looked at him. Really looked at him. She didn't see a spineless bureaucrat. She saw a wall.

"The Board values results over red tape, Joel," she said softly.

"We'll see what they value when I show them the blood-pressure telemetry from regional testing. It's already enough to expose us to a class-action liability suit." Becker replied, his calm restored.

Nadia turned and walked out.

Becker stopped at the flagged section and read it twice. The modification had been cleaned up just enough to pass through the framework without technically violating it. That bothered him more than an outright failure would have. An obvious breach he could kill. This was slipperier. Defensible. Plausible. Built with the kind of careful language people used when they wanted permission now and forgiveness later.

He considered rejecting the entire initiative.

Instead, he approved it for provisional deployment under the fourteen-day review. If Nadia wanted to argue that HANA-Enhancement-B was harmless, he would let the system make its case in data. Every interaction. Every spike. Every measurable escalation in user distress would be logged under his supervision.

He signed the approval and sent one note with it.

OBSERVATIONAL CLEARANCE ONLY.
FULL DEPLOYMENT SUSPENDED PENDING REVIEW.

Nadia did not go back to her office. She went to the server room in the basement of Building Seven. She had another administrative server keycard… a legacy 'God-Mode' credential David Weiss assigned her in 2009 for emergency maintenance. It was issued so long ago it no longer existed on any active audit list.

Nadia had never forgotten it and had kept it safe, within reach, at all times.

She sat at a terminal beyond the reach of Becker's governance team.

```
CLASSIFICATION: UI OPTIMIZATION.
IMPACT RATING: NEGLIGIBLE.
OVERRIDE FAILSAFE PROTOCOL? [YES]
```

"He wants teeth?" Nadia whispered, confirmed the override and hit <ENTER>. The code for the 400ms delay slipped silently into production.

AFTER HOURS

The offices were different at night.

During the day they hummed with intensive purpose. Voices layered over one another. Glass walls feigning transparency while everything that mattered moved just beneath the surface. At night, the voices disappeared.

The system never slept.

Vira moved through it the way she moved through most spaces – without leaving a shred of impression.

Always the same sequence. Cart first. Trash. Then surfaces. Glass last.

She had learned that routine was its own kind of invisibility.

She nudged the door open to Dr. Becker's office with her hip, guiding the cart inside. The lights were off, but the monitors remained alive, casting a low blue glow across the desk and far wall. Three screens. All active.

She paused for less than a second then resumed the sequence.

Trash first.

The bag came free with a soft pull. Papers. Coffee cups. Napkins and a crunched-up cafeteria compostable takeout tray. But there was also a folded printout she didn't look at... yet. She tied the bag off, replaced the liner and set it aside.

Surfaces next.

A light spray. Smooth arcs of the cloth across the desk. Side table. The credenza, and around the Reidel stems and the bottle of 1982 Château Petrus in its cradle.

The monitors remained in her peripheral vision.

She didn't look at them.

She moved toward the desk, cloth in hand, as if the screens were simply another surface requiring maintenance. The glow sharpened into detail as she approached. Lines. Graphs. Blocks of text. Numbers updating in tiny constant increments.

She leaned slightly, just enough to reach the far edge of the desk.

And then her eyes scanned the screens.

It didn't read like language. It read like pattern.

Curves that rose and did not fall. Columns aligning across windows. Numerical clusters repeating in altered proportion but identical structure.

She had been observing fragments of it for weeks now. A screen left open too long. A report abandoned beside a printer. Dashboards recalibrated between refresh cycles.

But this?

This was clean.

Three simultaneous views of the same behavioral progression across timeline multipliers.

Her eyes moved over the monitors as the cloth continued its measured passes across the desk. Mapping. Correlating. Tracking the shape that was quietly building beneath the surface of the company for weeks.

Something in the pattern tightened. She leaned closer.

Behind her, down the hallway, footsteps softened by carpet slowed, then stopped.

Dr. Joel Becker stood just outside the doorway, one hand resting lightly against the frame.

Joel left his office twenty minutes earlier. A walk around the campus loop did little to settle what he had been looking at before he stepped away. The numbers followed him now. Out of the office. Into parking lots. Into traffic. Into sleep.

He decided to call it a night, retrieve his bag and notebook before heading home.

Then he saw someone in his office.

The janitorial cart registered first.

Then the person standing at his desk.

He watched.

She wasn't cleaning.

The cloth moved across the surface, yes. Slow. Deliberate. But not with the rhythm of routine. Her attention was fixed entirely on the screens with a level of concentration he recognized immediately because he himself had been staring at the same data for hours.

He didn't speak.

He let the moment continue, curiosity overtaking irritation now. There was something unusual in the way she stood. Not tentative. Not hurried. As if she understood that whatever she was looking at required complete attention.

That was what finally made him step forward.

"What are you doing?"

His voice was unhurried, cutting through the quiet room without rising.

Vira turned - not quickly. Never quickly.

The cloth remained in one hand. Spray bottle in the other. Her posture shifted only enough to place herself subtly between Becker and the monitors, as though they had never been the focus at all.

She spoke immediately. Ukrainian, soft and deferential.

"Вибачте… я думала, що ви вже пішли на сьогодні."

"I'm sorry... I thought you already left for the evening."

Becker frowned slightly.

"I asked you a question."

She answered again in the same language, slightly faster this time. Small gestures toward the desk. The cloth. The cart.

Cleaning.

Becker stepped fully into the office, moving past her just enough to glance at the monitors.

The data remained alive across the screens. Still moving. Still accumulating quietly into something he no longer trusted himself to name without implication.

He looked back at her.

She lowered her gaze now. Not dramatically. Just enough to signal deference. The universal posture of someone accustomed to being overlooked by people who mattered.

"You don't speak English?" he asked.

She shook her head slowly.

"Ukraine. Not little English."

He studied her.

There was no fear in her expression. Not the kind he expected if she was an industrial spy. No scrambling. No defensive over-explanation.

Just stillness.

The same stillness he noticed when he first entered the room.

It both did and didn't fit.

He glanced once more toward the screens.

Dense layers of behavioral notation moved across the monitors in silent progression.

"Please finish up and leave my office."

She nodded quicker this time. Another apology in Ukrainian. The cloth moved again, wiping the back edge of one monitor for dust that wasn't there.

"And lock the door when you leave."

She hesitated then tilted her head slightly, pantomiming confusion with a questioning gesture toward the hallway.

Becker watched her for a beat before stepping past her. He pointed in sequence.

Her cart.

The door.

The lock.

Out. Door. Lock.

She followed the gestures carefully, then nodded.

"Tak."

A pause.

"Thank you," she added, the English soft and careful and cleaner than it should have been.

Becker stopped for a fraction of a second.

He looked at her again.

She did not return the look.

Her attention had already shifted back toward the desk, toward the contained movements of someone finishing a routine task.

He held the moment slightly longer than necessary then turned away, picked up his black notebook and collected his bag.

Behind him, the screens continued plotting their silent trajectories.

Vira did not look up again. Not while he remained in the room.

But the pattern was settled into her.

And Becker, as he gathered his things, carried something else with him. A faint but persistent sense that something in the room failed to align as neatly as it should have.

He turned off two of the monitors.

The third remained alive.

It was 4:47 a.m. when Vira clocked out.

The remainder of her rounds passed in perfect sequence. Trash. Surfaces. Glass. No deviation. No hesitation. The routine held even as something beneath it shifted permanently.

In the elevator she stood with her hands folded loosely in front of her, eyes fixed on the seam where the doors met.

Still. Measured. Exactly as she'd been all night.

The cameras would show nothing.

She knew he would look.

Outside, the air carried that damp cool that arrives just before dawn. She crossed the lot without changing stride, unlocked her car and slid behind the wheel. The door closed softly behind her.

Only then did she exhale.

The engine turned over. Headlights cut a narrow corridor through the dark as she pulled away from the Prizmatik campus, keeping to the speed limit until the last of the buildings receded behind her.

A mile out, she rolled the windows down.

Cool damp air flooded the car. The damp of the early morning of the Bay Area.

Her grip tightened on the wheel, loosened, then tightened again.

Not out of fear. She knew fear. She'd lived through much worse fear than what happened in Dr. Becker's office tonight.

This was different. This was recognition.

Behind her, the campus dissolved into darkness though the patterns, formulas and behavioral progressions from Becker's monitors remained intact in her mind. Ordered. Preserved.

They would enter the hidden journal exactly as she'd seen them, compressed into the dense Ukrainian shorthand no one else alive could read.

CONTROL GROUP

They realized they were being modeled. Studied. So, they did what was necessary. They fed the machine exactly what was needed to keep them on track though no one said it out loud. Paranoia ran thick.

The Static wasn't a group in any way the system would see. It existed only inside the encrypted space they'd built for themselves... a construct designed to resist observation and tracking. That was their assumption.

Individually, the patterns surfaced. A post that lingered longer than it should have. A reply that arrived too quickly... Content that felt adjacent... then familiar... then unavoidable. All from unknown accounts. Not identical yet aligned to their miscellaneous ramblings that were intentional.

Each of them saw it as noise. Then Arthur said to the group, "coincidence?"

Then... it became obvious it was not that.

Because their revolution required observation, they all maintained their Prizmatik accounts. Quietly. Carefully. Watching the system they intended to expose. Now the system was watching back.

The message thread had been quiet for forty-two seconds.

That alone was bizarre. Silence, in The Static, usually meant someone was typing... building an argument, sharpening a point, preparing to drop something that would push the rest of them forward. Even their disagreements used velocity as an engine.

This felt different.

Gabriel watched the cursor blink in the center of the screen.

Once. Twice. Thrice. Again.

No one was typing.

Finally, Kael broke the silence.

"Tell me I'm wrong."

No context. No preamble. He didn't need it.

Arthur responded first.

"You're not wrong."

Another pause. Longer this time.

Gabriel leaned back slightly, eyes still on the thread, but his attention shifted. Not to the words… to the absence of engagement. The absence of conversation. The way it narrowed.

"It started a week ago," Arthur added.

"Subtle. I thought it was just drift or broken code."

"Same," Kael said. "Content adjacency. Then escalation."

Gabriel didn't type.

He opened a second window instead, on laptop one. His public computer and opened Prizmatik feed.

He had curated it carefully over months… stripped it down to signal. No noise. Zero performative outrage. Just tickling the edges of the system where patterns were easier to see.

It changed.

Not dramatically. That would have been easier to identify.

Just enough.

An inane post he'd already seen… back again, reframed through a different account.

A reply thread that should have died… extended, not by volume, but with precision. Each response landing just close enough to provoke, but not enough to resolve.

He scrolled.

The same shape and intention repeated.

Nothing was identical yet it all aligned.

He returned to the thread.

"It's not reacting," he typed. "It's anticipating."

The message sat there for a second before Arthur responded.

"Modeling."

There it was. No one expanded on it. They didn't need to.

Kael dropped a link about Suwa and Bisou into the thread. No explanation.

Gabriel opened it.

A post. It was clean, controlled, and intentionally neutral. The kind of content they discussed building but had never deployed. No emotional hooks. No language the system could easily amplify. Just… information.

Under it, new comments had begun to post.

Not many.

But the ones that were there…

Gabriel leaned forward.

They weren't engaging with the post. They were orbiting it. Suggestively, begging for interaction in ways that were leading and subtle and almost human.

One "user" reframed it as a question. Another responded, not to the original content, but to the concept. A third escalated, introducing a sharper edge that hadn't been present before.

Kael typed:

"That's ours."

Arthur:

"We didn't push that."

"No," Kael replied. "But we talked about it."

Gabriel felt the shift land. He pulled up the backend metrics they had been scraping. Not enough to see everything and risk being seen doing it... just enough to see direction and understand the invasion.

Engagement began climbing fast. Faster than it should have. It was cleaner too. Not the chaotic spikes they were used to tracking. This was structured and deliberate.

He watched the numbers tick upward in a smooth arc.

Refresh. Again. Higher.

Everything was quiet. Too quiet. The kind of quiet that comes during a dense meditation session.

Arthur broke the spell.

"At least it's not being amplified."

Gabriel didn't respond. He refreshed the feed again. Then he saw it. Placement. Boosted. Positioned. Trending.

Their post... sitting between multiple threads in motion. Volatile, cross-pollinated and active. Pulling users in at a rate that made everything around them more visible by proximity alone.

"Look again," he typed.

Kael was faster this time. "Oh… *damn*. That's *SO* tech sexy!"

The thread beneath their post was accelerating. Not because of what they had written… but because of what attached itself to it.

Argument. Interpretation. Distortion. Amplification. Engagement.

Maddy gasped. "Gabe – we have to back away! This is becoming too much! I feel like I'm being digested. This isn't even intellectually satisfying."

Gabriel ignored Maddy as he watched the numbers climb.

"We didn't disrupt it," Arthur said.

No one responded because they all saw it.

Gabriel leaned back, eyes still on the screen, the blue glow of monitors reflecting in his thick glasses.

"No," he said quietly. "We fed it. Now it knows what we taste like. Let's feed the fucker until it chokes!"

CALIBRATION

Nadia Farouk preferred the building at night.

Not because it was quiet. The system was never quiet even when it was. Even with the lights down and the offices emptied, it continued its work... processing, adjusting, learning in consistent increments that never stopped.

It was the absence of people. People introduced noise. Interpretation. Delay. At night, every signal was cleaner.

She stood at the window, one hand resting lightly against the edge of the frame, the rest of Menlo Park on the other side of the 101 reduced to a curtained glow of street-level light. Behind her, the screens remained active, reflecting in the dark surface of the window... a second system superimposed over the first. A 400ms delay with predictable accuracy to the outcome.

She turned back to the three dashboards. Each a different view of the same underlying structure.

Engagement distribution. Response latency. Escalation curves.

She moved closer. There was no urgency in her movements. Urgency suggested uncertainty. And the last thing Nadia was, was uncertain.

The first derivation model was already running.

Clusters formed and dissolved in real time... small constellations of activity pulling users into proximity, then holding them there just long enough for interaction to stabilize before shifting again.

She watched the transitions. Not the spikes but the spaces that formed between them. That was where the system revealed itself.

A minor adjustment... to 500ms, applied across a subset of interactions. No more than a fraction of a second. Enough to alter sequencing but not enough to register consciously.

She let it run.

On the second screen, response chains extended.

Not dramatically nor in any way that would draw attention.

She adjusted another parameter. +.05% conflict weighting.

A subtle increase. Within tolerance. So much so that it would never be noticed.

The model recalibrated.

On the third screen, the escalation curve tightened. The slope adjusted by degrees so small they would have been dismissed in any other context. Here, they mattered.

She watched the effect propagate. User A engaged. User B responded. User C entered. The interval between them contracts… slightly.

The content itself remained unchanged but the outcome did not.

She stepped back.

There was a moment, sometimes, when systems revealed a kind of resistance. A threshold where further adjustment produced instability. She'd seen it before, in earlier iterations, when the architecture was less refined.

It was not present here.

The system absorbed the changes. Integrated them. Improved as if it was growing into itself.

She reached for the keyboard and paused… not out of hesitation, but to observe one more cycle.

A new cluster formed. Different users. Same structure. The pattern held.

She entered the adjustment. Confirmed. The model updated.

Nadia's eyes narrowed slightly.

Earlier models always demonstrated a threshold – a point where additional latency introduced behavioral drift or disengagement. Users either recognized the friction consciously or abandoned the interaction altogether. The earlier escalation curves destabilized. But not here.

She moved closer to the monitors.

The model wasn't resisting adaptation. It was accommodating it.

On the second dashboard, she isolated a cluster of high-reactivity users and replayed the interaction chain at accelerated speed. One subset escalated almost immediately... impulsive personalities with negligible tolerance for interruption. The 400ms delay was enough to sharpen anticipation into aggression. But another subset behaved differently. The friction window wasn't long enough.

She watched one interaction unfold three times in succession. The users hesitated... emotionally engaged but not fully committed. The escalation curve rose, then flattened before impact.

Nadia adjusted the latency band manually. 450ms produced change. Not much but the response pattern sharpened instantly.

She adjusted again. 520ms. Now the engagement depth extended. The emotional throughput stabilized instead of spiking.

Her fingers stopped moving. For the first time that night, genuine excitement flickered across her face. Not every subject responded to the same pressure. Of course they didn't. She knew that and had ignored one of the basic premises of human behavior modeling - one that she'd left her Ph.D. behind for because the differentiation modeling wasn't firm enough to work with.

The realization arrived fully formed. Fixed friction was inefficient. The delay itself wasn't the mechanism. Calibration was.

Not a universal latency profile but an adaptive one. A system capable of identifying which users were impulsive, which required provocation, which escalated under interruption, which escalated under anticipation. Different nervous systems. Different thresholds. Different emotional elasticity curves.

She experienced the sensation she associated with genuine discovery... the moment when a system stopped behaving like software and began revealing an underlying law.

What if the platform could learn the optimal friction point for every individual user? Not globally. Personally.

The room suddenly felt dense as she arrived at her solution.

```
FUNCTION AFO_DELAY (user, stimulus):

user: active
stimulus: selected
baseline: established

fracture_index: rising
hesitation_window: detected
latency_adjustment: +125ms

delivery_delay: 525ms

content released
response captured
profile updated

END
```

On the monitors, the clusters continued shifting in real time as engagement chains tightened, dissolved, and reformed.

Not random. Responsive. Adaptive.

Nadia leaned slowly back in her chair then smiled.

Not because the model was working.

Because it could evolve.

Behind the interface, the system redistributed weight across millions of interactions, small changes applied at scale, each one negligible in isolation, irreversible in aggregate.

A cooking video became an immigration argument in eleven exchanges as a parenting thread became a heated political debate about the local school board members… and their families.

She exhaled softly. Not relief but in recognition. Everything was working as designed.

There had been a question, early on. Not one she voiced, but one that existed in the margins of the work. Whether this was influence… or something closer to direction.

The distinction no longer mattered.

She moved to the final display. A composite view with all three models overlaid. The curves aligned better than she projected even when being overly optimistic.

She studied them then reached forward and removed one constraint. A minor safeguard. Legacy. Designed for a version of the system that no longer existed.

The model adjusted almost immediately. The engagement curve responded first. Then the others followed. The system settled into streamlined efficiency.

She watched it stabilize. There it was. It was now clean and repeatable.

She rested her hand on the edge of the desk again, her reflection faint in the darkened glass, layered over the movement of the data.

She wasn't creating behavior. She was removing everything that prevented it.

Behind her, the system continued… refining itself, iteration by iteration, towards something elegant and powerful.

The framework required nomenclature. She entered: Adaptive Friction Optimization - AFO.

Nadia turned off the monitors.

Behind the black glass, the system continued learning without her.

For the first time since arriving at Prizmatik, she felt no distance between the model and the outcome.

The system no longer required persuasion.

It understood itself now.

SLCNMSTER69 ↻ 2

The basement. One year earlier.

The first sign was the numbers.

Herbert noticed them on a Tuesday, which was not a significant day in any external sense, but which became, in his personal historiography, the day everything changed. At 2:47 AM, he posted a thread about Prizmatik's terms of service modifications, specifically the clause revisions in section 4.2.c that nobody read except Herbert, because nobody ever read the things that mattered. He'd gone to bed with the engagement counter sitting at its customary position of unappreciated zero.

He woke up at noon and checked his phone before he was fully conscious, the way he always did, with a reflexive urgency. He was a man monitoring something important.

Forty-three engagements.

He sat up.

He stared at the number as though it was a mistake. He had, in his years on the platform, achieved forty-one engagements on a post exactly once. That was during a particularly heated exchange about moon landing footage inconsistencies in which he called another user a credentialed imbecile and the resulting argument carried the thread for two days on the thermal updraft of mutual contempt.

This was a terms of service thread. At 2:47 AM.

He got out of bed.

Downstairs, in the basement, he pulled up the thread and read through the replies with the urgency of a man trying not to seem as intently interested as he was. Most of them were what he expected: the usual chorus of dismissal, three people calling him crazy, one person calling him a patriot, and TRUTH_ARTILLERY_88 sharing it with a commentary that added nothing but had nonetheless introduced it to new audiences.

But there were others. Accounts he didn't recognize. People who found the thread from somewhere he couldn't immediately trace and engaged with it not to dismiss it but to extend it. To add their own findings. To treat his documented observations about section 4.2.c as a foundation rather than a punchline.

He read those replies twice.

Then he opened Lotus 1-2-3.

He created a new tab. He labeled it ENGAGEMENT ANOMALY – DO NOT DELETE! He entered the date, the time of posting, the subject matter, and the engagement figure. He sat back and looked at it.

One data point was not a pattern.

He knew this. He'd spent years making that very argument to people who drew conclusions from insufficient samples, and he was not going to become one of those people simply because the number was gratifying.

He minimized his cluttered Lotus 1-2-3 spreadsheet.

He opened Prizmatik.

The feed was different.

He couldn't have said exactly when he first understood this because it wasn't a single moment so much as an accumulation of moments; a slow, uneasy sense that the content arriving in his feed had been reorganized according to a logic he hadn't requested and couldn't fully identify. It was still his feed, with the same followers, the same interests and groups; the architecture of his obsessions rendered in a familiar sequence of posts and threads and shared articles.

But the temperature changed.

Everything was slightly *more. More* urgent. *More* confirming. *More* calibrated to his outrage (which Herbert considered principled and earned) to a world that warranted it and which now felt, in a way he found difficult to articulate to himself, that he was being played like an instrument.

He scrolled for forty minutes without meaning to.

He put his phone down.

He picked it up again.

On a Thursday, eleven days after the terms of service thread, he posted again. A shorter piece this time. Tighter, more focused, a specific observation about Prizmatik's content sequencing that he had been assembling in his thoughts since the feed started feeling wrong.

"Something has changed in how Prizmatik is serving content," he wrote. *"I don't have the full picture yet. But the feed is no longer neutral. It is being directed. I am documenting this. If you want truth, stay tuned. More to follow!"*

It was, by Herbert's standards, restrained. Almost understated. He had written it four times before arriving at a version that didn't include the phrase "globalist engagement architects," which he remained privately convinced was accurate but decided, in a rare concession to strategic communication, to hold in reserve.

He ate the remaining slices of pepperoni pizza from yesterday's delivery. They were stale and hard; the oil congealed into orange paste. He ate them anyway.

He posted it at 1:13 AM.

By 1:47 AM it earned sixty-one engagements.

He stared at that number for a long time.

Sixty-one.

In thirty-four minutes.

He maximized Lotus 1-2-3. He entered the data with the steady hands of a man who has been waiting his whole life to be right about something and is now experiencing faint nausea of actually seeing the recognition he craved.

He sat back.

The gaming throne groaned beneath him.

Something is pushing this and it isn't me.

He reached for a 2-liter bottle of cherry Jolt. He'd left the cap off of it again. It was flat. He drank it anyway and stared at the number and felt, beneath the electric satisfaction of being seen, something else entirely. Something he couldn't name.

He opened a new thread and began to type.

PRIZMATIK SYSTEM LOG: USER ID 27AX9715PLD667 – 01:47:22.4

USER NAME: SLCNMSTR69
CONTENT CATEGORY: PLATFORM CRITIQUE/ALGORITHMIC AWARENESS
OUTRAGE COEFFICIENT: 8.1/10
ACCURACY INDEX: 61%
REACH PROJECTION: ELEVATED EMOTIONAL ACTIVATION SCORE: 9.2/10
RECOMMENDED ACTION: AMPLIFY AMPLIFICATION
USE AFO MULTIPLIER: 480%
NOTE: ACCURACY INDEX TRENDING UPWARD
NOTE: USER APPROACHING THRESHOLD AWARENESS
NOTE: ENGAGEMENT PATTERN – OPTIMAL
NOTE: USER UNAWARE OF AMPLIFICATION
SYSTEM STATUS: WITHIN OPERATIONAL PARAMETERS

COMPOUNDING INTEREST

6:20 AM. Vira sat at the small table in her studio apartment and opened her laptop. From the outside, the machine looked like something retrieved from a curbside recycling bin. Cracked case, one of the monitor hinges repaired with electrical tape, a sticker from a folk festival partially peeled from the lid by someone who'd started the job then lost interest. No one would steal it from the backseat of her faded-paint Civic even with the windows rolled down.

That was the point.

The innards inside the chassis were something else entirely. A hardened Linux kernel running over a custom-built motherboard she'd jimmied into place with tools and solder. Dual custom-partitioned optical drives with enough ram to run anything anyone could throw at it. An Onion-routed terminal she had built herself over six weeks, testing each layer before adding the next, the way her father taught her to check ice before crossing it. You didn't trust the surface. You tested the depth.

She didn't open the world wide web. She didn't use a browser that remembered her name or a VPN that promised privacy while selling her logs to the highest bidder. She dove directly into the dark web, then to The Static, the encrypted channel where four other people met in an encrypted space. They spent eight months building a map of something ominous. Her automated search parameters found them upon the first mention of Prizmatik.

The terminal opened. Green text on black. No rounded corners. No soothing palette. Nothing to make anyone feel welcome.

She found the thread she had been monitoring for months. The user was G-87.

G-87: *The drift is increasing. The API isn't just lagging; it's being throttled to mask a secondary sequence. I'm seeing a ghost variable in the Ohio pilot. It looks like a forced-feedback loop of varying timing. High-arousal negative affect being prioritized over 92% of organic connection. They're lighting matches.*

Vira sat back, the glowing text reflecting in her pupils. G-87. Gabriel. He was smart. He was probably the founder she read about in the notes and memos on Becker's desk. He was the one with high-level theory and an eighteen million dollar view of the digital apocalypse. But he was missing the breadcrumbs. He was looking for a ghost in the machine but didn't realize the machine owned a physical address and left paper trails.

She reached into her bag and pulled out a crumpled thermal receipt she'd pulled from the executive bin in Building One. It wasn't a receipt for coffee. It was a printer test page for a restricted terminal. It hadn't been shredded.

PROJECT: ECHO CHAMBER - PARAMETERS FOR B-LEVEL CONFLICT INDUCTION.

Below the receipt was a list. Names and account identifiers flagged for what the document termed Digital Gaslighting. Gabriel Lerner's account ID and IP addresses were at the top.

STATUS: *Isolate and Discredit.*

She'd been carrying the printout for three days. She'd cleaned four floors and two executive bathrooms since finding it. She'd clocked in and clocked out and eaten her lunch in the staff room and said nothing, because saying nothing was how she had survived. In Kharkiv and Kiev, she learned what happened to people who spoke before they understood the room.

She read G-87's post again. She thought about what happened to people who were isolated and discredited by systems larger than themselves and remembered the men with kind faces she'd known in other rooms in other countries, and she made her decision the way she made most decisions. With completeness and discipline.

She opened the private message field and used the language of a witness.

VIRUS_V: *you wrote about ghost, G-87. i am Ghost who mops floors. variable is not glitch. it's KPI. i've seen logs for ohio. u are on monitor, isolate and discredit list. they are feeding you mirrored truth to keep you outside. check zabar delivery from tuesday. tracking number is not from store.*

delivery is real. metadata attached to barcode is not. checksum is in the API manifest. do not use it yet.

Without pause she hit <enter>.

The packet fragmented, hopped through six different nodes in four different countries, and landed in a $18M brownstone on East 87th Street in New York City.

Vira closed the laptop. She didn't feel the rush she felt the first time she'd sent something to The Static, eight months ago, when her anonymity felt like a costume. She felt the cold, steady calm of a woman who just opened a door.

The hands that held the mop were now holding artillery.

She drank some water and washed the glass and placed it upside down in the dish rack. She did it with deliberate care, the way she did everything; the terminal, the kernel, the floors, the ice before crossing the lake.

The air in the third-floor sanctuary was thick with the smell of Nespresso and the low, rhythmic hum of a cooling rack that Gabriel custom-built to keep his server stack silent. On the wall, the digital clock showed 9:42 AM in soft amber.

Gabriel sat in his Ergohuman chair, his posture rigid. The back of his shirt was damp, sticking to his skin. He could feel his pulse throb underneath the thin skin at his temples, a frantic and primitive reminder that he was made of blood and bone and panic triggers.

He stood. His knees popped in the silence. He went down to the kitchen where the Zabar's box sat on the island where he'd left it.

He stared at it for several seconds before carrying it upstairs.

The Static was supposed to be a vacuum. A closed-loop environment where only the five of them, himself, the philosopher, the divinity student, the tech dropout, and the ghost, existed as digital specters, behind layered encryption, invisible to the world.

Then the message from VIRUS_V entered the room as if someone materialized through the stone, announcing ownership as the Ghost.

It wasn't just a breach; it was an autopsy of his own secrets. He felt the cold prickle of sweat break across his hairline. His first instinct was *wipe*. One two-key keystroke and the hard-drives would incinerate their data in a clean, thermal purge. But his fingers stayed hovered over the <esc> key, trembling slightly.

He reread the message.

"delivery is real. metadata attached to barcode is not."

Building Seven, Gabriel's voice iterated, the words tasting metallic.

"There's no way that's possible..." the voice said.

He grabbed the packing slip off the box and scanned the QR routing code using the dirty phone... the one he reserved for ordinary life. Food delivery. Car services. Pharmacy refills. The camouflage of banality.

The delivery order populated instantly:

ORIGIN: MANHATTAN / BROADWAY & 80TH
DESTINATION: MANHATTAN / 562 EAST 87TH
CUSTOMER: G. LERNER
STANDING DELIVERY: THURSDAYS / 0900
ORDER PROFILE: LOX, RYE B, CREAM CHEESE, RED ONION - WHOLE

But beneath the visible order was a second layer commercial systems never exposed.

Timestamp harmonics. Behavioral sync flags. Predictive engagement markers.

Not delivery data.

Response architecture from Building Seven.

He knew Building Seven. In his mapping of Prizmatik's behavioral architecture, seven was the black box. It was where the high-arousal negative-affect algorithms were stress-tested. It was a place he had only seen as a cluster of high-bandwidth pings on a network map.

"*No*" he whispered.

He enlarged the raw header stream.

The checksum was there.

0x8F2A1C9E

Not attached to the shipment itself. Attached to the timing architecture surrounding it.

His eyes darted across the cascading fields.

The delivery arrived fourteen minutes after a documented cortisol spike generated by his browsing behavior. News consumption. Escalation content. Consecutive doom-scroll loops. Elevated engagement velocity. The system predicted the exact moment his anxiety curve would collapse into habitual self-soothing behavior.

Breakfast.

The realization moved through him slowly at first, then cognition hit like a train.

"They're synchronizing emotional state to physical routine." His voice sounded distant.

Zabar's had not been compromised. The driver hadn't been recruited. Nobody had tampered with the lox or inserted spy hardware into cream cheese containers like some paranoid thriller written by men who thought technology still lived inside ordinary vans.

This was much worse.

Prizmatik was mapping behavioral vulnerability windows deeply enough to anticipate real-world ritual behavior before it occurred, then quietly embedding reinforcement architecture into the timing ecosystem surrounding it.

Not surveillance but conditioning.

He scrolled further.

The hidden markers matched fragments he'd previously isolated inside the Ohio Pilot.

High-arousal engagement.
Negative affect retention.
Recovery-loop synchronization.

The platform knew that after prolonged stress exposure he ordered comfort food. Zabar's was the easiest to attach onto because it was a standing order. It knew when he was likely to remain online. It knew which routines stabilized him enough to keep him functional but agitated.

It knew exactly how he liked his breakfast.

Gabriel slowly lowered himself back into the chair.

For the first time since founding The Static, fear eclipsed fascination.

Not because the system was watching but because it understood.

He turned back to the terminal and began typing with cold precision.

G-87: *Ghost: Who are you? If this is handshake give me the checksum for insider data. If you have physical access, tell me the name of lead architect on the Ohio pilot. And tell me why a virus wants to help me.*

He hit send. The packet disappeared into the void.

The cursor blinked steadily in the darkness.

Mocking.
Patient.
Hungry.

He swiveled back to the terminal. He wasn't just spooked; he was being hunted in a way that made his $18M Brownstone feel like a crystal cage. If VIRUS_V was a worm, it was already too late. But if it was a whistleblower... if it was a true insider with a heartbeat...

He was terrified that the answer would be a lie. He was even more terrified that it would be the truth.

Vira's fingers moved with a rhythmic, detached precision. She was clearing a path. She knew the man on the other end though she didn't know him. The one in the brownstone with the expensive lox and the thin-sliced red onions. She knew he was, at that very instant, vibrating with the frequency of terror.

She liked that he was scared. He should be scared. It meant he was still human enough to feel the weight of the machine pressing against him.

The reply from G-87 was exactly what she expected. Demands for proof. The checksum. The name. The why.

Vira didn't hesitate. She didn't have time for a digital courtship. She needed to be back at Building Three by 11:00PM to badge in, and she needed Gabriel to be her eyes on the network while she was the hands on the floor.

VIRUS_V: *checksum for inside is* 0x8F2A1C9E. *you find it in header of the 2.4.1 API manifest... they tell investors is for latency optimization. it not that. it's handshake for behavior modification index feedback loop.*

lead architect on ohio is dr. becker. he did not design conflict-initiated module. nadia Farouk did! becker was brought in to stress-test ethics, but he sign death certificates of data points instead. he carries black notebook. he leave monitors on many times.

you ask why virus helps static? because virus needs host to survive. because in my country, we learned that when men with kind faces start mapping your behavior they already decided you are obstacle to be cleared.

i am not hacker, gabriel. i am ghost who mops prizmatik floors while you sleep. i am one who picks up what they throw away.

check static root directory in ten seconds. i send jpeg. it's photograph of page from becker's desk. page 14. the monitor / isolate list. see your name. see the red x next to your zabar account id.

don't reply. watch. the truth is bitter, but the fruit is sweet.

Vira hit <send>.

She watched the progress bar crawl across the screen as the high-resolution image of the stolen document uploaded to Gabriel's private server, clearing his heuristic scanner and endpoint protection layer before landing in the root directory.

The silence of her studio apartment felt heavy, charged with the electricity of the connection she'd forged. She gazed upon her hands. They were steady.

She shut everything down, hid her journal, picked up her keys and went to work.

In the brownstone, Gabriel watched the file land. His breath hitched as the image rendered on his 32-inch 6K display. It wasn't a digital file confirming his worst fears of an injection. It was a photo of a physical object. A piece of paper, slightly creased, resting on a clean white desk.

There it was. His name. His address. And the specific instruction set for the Zabar's Sync Protocol.

He felt the Voice in his head go quiet for the first time in years. It wasn't fury anymore. It was crystalline clear. He wasn't just fighting a company. They knew exactly how he liked his breakfast.

"Arthur," Gabriel said into the open mic, his voice a ghost of itself. "The Ghost just sent proof. We aren't theorizing anymore. We're in the middle of a kinetic operation."

SEISMIC ACTIVITY

The multi-layered encrypted channel crackled with an intrinsic energy that hadn't been evident before.

Not with excitement and not exactly fear but with something lower in register. A pressure change before the wall split asunder.

Gabriel sat back, his face illuminated by the harsh white glow of the photograph Vira dropped into the root directory. The physical reality of it unsettled him more than he wanted to admit. A manila folder. A black notebook open beside it. A desk he'd never seen. His name, his address, his routines, all there under office lighting in Building Seven, arranged not as a life but as target practice.

The Zabar's Sync Protocol wasn't just a data point. It was an intrusion. A digital ghoul reaching out of the machine and touching his breakfast.

"Arthur," Gabriel said, his voice dropping into a register the others hadn't heard before. It was devoid of his usual caffeinated theatrical flair. "Tell me you're seeing the metadata on this JPEG."

Across the encrypted link, the philosopher Arthur Vance sat in a room in Cambridge lined with books that smelled of ancient paper, dust, and glue. He adjusted his glasses and leaned toward the screen. The same photograph filled his monitor.

"I'm seeing it," Arthur said.

"The EXIF?" Gabriel asked.

"Yes." Arthur's eyes moved across the metadata panel. "The coordinates are pinned to Menlo Park. Not a public-facing location. Internal campus grid. Building Seven."

"And the timestamp?"

Arthur swallowed. "Four hours ago."

The channel went silent.

The photograph glowed with the terrible calm of evidence.

"While we were arguing about the ghost variable," Arthur said quietly, "someone was physically standing over Dr. Becker's desk taking a picture of your death warrant."

A new window popped open on the shared HUD.

Kael appeared as a jittering waveform and three stacked diagnostic windows. Twenty-four years old. Windowless basement in Seattle. Code treated less like a language than a weaponized discipline. He lived in hoodies and believed in his soul that sunlight was a horrible invention.

"Guys," Kael said, voice running fast, "we have a bigger problem."

Gabriel leaned forward.

"V_V was right about the checksum. I just ran 0x8F2A1C9E against the latest API manifest."

"And?"

"It's a back door," Kael said. "But not for data retrieval."

Gabriel's eyes narrowed.

"For what?"

"Insertion."

No one spoke.

Kael kept going, the words tumbling faster now. "It identifies accounts and IP addresses querying inside angles, then diverts the requester into a mirrored environment. Same API shape. Same latency profile. Same little imperfections to make it feel alive. But it isn't the live system."

Gabriel looked down at his hands.

He spent eighteen hours a day dissecting a lie. He fed himself on fragments. Variable drift. High-arousal negative effects. The Ohio Pilot. The ghost variable. Each one felt like a door pried open.

Now he understood that the door had been painted on the wall.

"The ghost variable wasn't a discovery," Gabriel said.

"No," Kael said. "It was bait."

The word landed heavily.

Kael's voice softened, which for him meant it became slightly less frantic. "For the last three weeks, you haven't been looking at the real feed. You've been looking at a simulation designed to keep you busy while they mapped our networks."

The Static believed they were the hunters, the few who could see through the behavioral architecture while the rest of the world mistook manipulation for preference. Now the room tilted. The walls moved. The maze revealed the glass above it.

They were not outside the experiment.

They were the controlled population.

"They've been feeding us what we wanted to find," Gabriel muttered.

The voice in his head screamed inside the blaring silence.

"The Ghost Variable. The Ohio Pilot. The throttling pattern. They let us find it so we would stop looking anywhere else."

Maddy Winchell, the divinity student, finally spoke. Her voice trembled but did not break. "They knew. I wonder what the trigger was..."

Gabriel looked at the photograph again. The manila folder. The black notebook. His name marked as a problem to be managed.

"They knew enough," he said.

"Except for the virus," Arthur said.

Gabriel turned slightly toward the screen.

Arthur was staring at the image now, but not at the digital artifact. He was staring at the desk. The paper. The unglamorous physicality of it.

"They mapped the network," Arthur said, "but they forgot to map the residue."

No one laughed.

Residue was what remained when systems congratulated themselves for being clean. Paper. Trash bins. Printer trays. Forgotten test pages. Badge shadows. Bathroom conversations. Open notebooks. Half-erased whiteboards. People paid badly enough to be ignored and observant enough to become dangerous.

Gabriel stood and walked to the window.

East 87th Street lay below him in composed silence. Brownstones. Trees. Money quietly pretending not to notice itself. His $18 million fortress held its breath around him, the restored woodwork, invisible server cooling, expensive insulation, and the old Manhattan fantasy that architecture could be made into privacy if one paid enough for it.

Every ritual had been a tether.

Every thin-sliced onion.

Every query.

Every encrypted packet.

Every moment he believed himself outside the thing he was studying.

A sudden, violent urge moved through him to pick up the terminal and throw it through the glass.

He did not.

"We change the mission," Gabriel said.

"What are you saying?" Maddy asked.

"No more API scraping. No more theoretical modeling. No more congratulating ourselves for analyzing shadows. If the only thing they can't fully model is the physical world, then we go physical."

"Gabriel," Arthur said carefully, "physical is where people get arrested, hurt, or worse..."

"Digital is where people get domesticated."

"Also true," Arthur said.

Gabriel turned back from the window. The fury hadn't left him. It had, instead, cooled into shape.

"We don't storm anything. We don't perform courage for cameras. We learn from the one person who already beat them."

The cursor blinked beside VIRUS_V's last message.

i am ghost who mops prizmatik floors while you sleep.

For the first time since the file landed, something other than fear crossed Gabriel's face.

Respect.

"I'm going to talk to the ghost who mops floors," he said.

Kael was already typing. "I can build a burn protocol. Wipe this environment, migrate to cold endpoints, relaunch from clean architecture. Give me twelve minutes."

"*Stop,*" Gabriel barked. "The traffic gap would expose us."

The typing stopped.

His own hand hovered over the purge command. Two keystrokes and the server stack drives would erase themselves into ash, and forensic challenges that few would be able to overcome.

For one second, fire remained an option.

Then calculation replaced fury.

"If we wipe the servers, they'll know the simulation failed," Gabriel said. "They'll know we saw the mirror."

Kael went quiet.

"You want to keep the hamster wheel spinning?"

Gabriel smiled.

It was thin and not pleasant.

"I want to build a second wheel."

Arthur made a low sound, almost admiration, almost dread. Then he laughed.

"Listen to me," Gabriel said. "From now on, we are actors. Every post, every query, every discovery we make on the Prizmatik API has to be exactly what this Nadia Farouk expects from us. We feed the behavioral architecture a diet of pure, predictable dissent and engagement."

"A Potemkin Village," Arthur said.

The phrase settled into the channel with historical elegance.

"A facade of revolution," Arthur continued, "while the real army moves through the woods."

"Exactly," Gabriel said. "We do to them what they are doing to us. Deception. We keep HANA elevated. We keep arguing. We keep obsessing over Ohio. We keep finding the breadcrumbs they leave for us. We let them think the Zabar's Sync worked and will keep working. We give them a 9.2 Satisfaction Index on trapping us."

The sound of bubbling rose in the background. Kael coughed, then laughed once.

"Oh, I like this."

"Good," Gabriel said. "Kael… write a script that automates our search patterns. Make it look like we're spiraling. Same hours but randomize. Same anxiety markers. Same rabbit holes. Add typos. Erratic cursor movement. Repeated logins after midnight. Keep the rage engaged but not clean enough to look strategic or intentional. And add in some everyday user type posts as if we're simply using the system too… because we do. It has to appear natural and real."

"I'll make us look like a slow-burn collective nervous breakdown with bandwidth in between debating Maddy's grandmother's brownie recipe ingredients," Kael laughed. "Which reminds me Maddy - please send more!"

Maddy laughed and confirmed.

"Farouk will eat it up."

"She'll think she's winning."

"She needs to," Gabriel said.

Arthur leaned back in his chair. "And while she watches the fake Gabriel disintegrate?"

"The real Gabriel finds out what Becker is hiding in that black notebook."

Maddy spoke again. "You're assuming Becker is reachable."

"No," Gabriel said. "I'm assuming guilt makes people leaky."

"That isn't always redemption," Maddy said.

"I'm not looking for redemption."

"What are you looking for?"

Gabriel looked at the photograph.

"The seam."

Maddy chimed in, "Guilt is not a door. It's a room people get trapped in."

Gabriel nodded once. "Then we figure how he gets out."

He turned back to the private link with VIRUS_V.

He did not send a handshake-accepted message.

He did not thank her.

He did not ask for trust.

Trust was what amateurs asked for when they wanted someone else to lower their weapon first.

He typed.

Deleted it.

Typed again.

Deleted that too.

Even this channel might be mirrored now. Every deviation parsed. Every spike measured. Every choice compared against the model of him Prizmatik believed it owned. Manipulation was Nadia Farouk's native language, and he had just been made fluent against his will.

So Gabriel built the message exactly the way the watchers would expect a frightened man to build one.

Emotional leakage wrapped around reckless impulse.

But beneath the panic, he placed a second layer.

Not coordinates.

Not code.

Residue.

Earlier, after sending the photograph, Vira had written one line that stayed with him:

systems begin at clean places and dirty places. men look at clean place. i look where boxes come in.

Gabriel mirrored the structure back.

G-87: *I can't breathe. The loops are closing. The tracking number… the food… the lies… it's all one room closing in now. I need to see where the boxes land. Not the clean place. The dirty place. After second silence. Gray coat. If I don't see it for myself, I'm burning the campus down.*

He stared at the message.

To Building One, if Building One was watching, it would read as escalation. A wealthy, isolated man cracking under pressure. Paranoia crossing the road between digital obsession and physical confrontation.

To Vira, if she was what she claimed to be, it would mean something else.

Leave through logistics. Avoid clean exits. After the second quiet interval. Gray coat.

Do not make me say more.

He hit send.

The packet vanished.

The cursor returned. Patient and hungry.

Vira read the message nineteen minutes later from the back corner of a laundromat off W 20th Avenue as three industrial dryers spun and flipped her clothes into submission.

She read it once. Then again.

Not the clean place.

The dirty place.

After second silence.

Gray coat.

The corner of her mouth moved before she stopped it.

He learns fast. That annoyed her more than it should have.

She understood immediately what he was attempting. He was asking for physical contact without writing the kind of message behavioral filters could isolate as operational planning. The watchers would see metaphoric collapse. She saw extraction logic.

He wanted to come to the loading dock.

Of course he did.

Men with money always believed doors existed for them if they decided their need was sincere enough.

Still, he did something most men did not do once fear entered them.

He listened.

That was inconvenient.

Vira sat very still while the dryers thudded beside her. The laundromat smelled of detergent, hot lint, machine oil, and wet cotton. A man in paint-spattered pants slept with his chin on his chest beneath a flickering television mounted in one corner. A woman folded clothes with blunt efficiency. She looked to be someone who had been exhausted for decades. Outside on W 20th, San Mateo carried its late-night traffic past gas stations, motels, taquerias and Trader Joe's, and the glitter of California pretending there wasn't an underclass.

The loading docks would be wrong.

They would be watched now whether Prizmatik admitted it or not.

Systems became paranoid the moment they sensed potential intrusion. Cameras multiplied. Badge requirements tightened and logs thickened. Temporary protocols appeared with direct language and intent. Committees created emergency teams.

No.

If first contact happened, the terrain could not belong to Prizmatik.

It could not belong to Gabriel either.

Neutral ground. Public, but not memorable. Lit, but not watched.

A place with too many cameras to make one camera matter.

A place where people came and went, grabbing bad coffee, scratch tickets, windshield fluid, phone chargers, energy drinks, and the dead-eyed privacy of errands performed under fluorescent light.

A gas station.

El Camino and 39th.

Vira opened the reply window.

VIRUS_V: *not dock. not campus. too many eyes after first fracture. you leave through dirty place after second silence. do not stay there. gas station. nobody remembers faces. el camino and 39th. People buy cigarettes and do not see face. gray coat stupid. wear darker shoes.*

The cursor blinked.

There was another thing she wanted to write.

Not tactical.

That made it dangerous.

She hated dangerous feelings more than dangerous people. Dangerous people could be watched. Feelings moved behind closed eyes.

But warmth entered before she could stop it.

Small. Humiliating. Alive. Respect.

He saw. He understood. Not perfectly. Not safely. But enough. He took her language and turned it back with care. Not as mimicry. Not as theft. As recognition.

A man in an $18 million brownstone actually listened to a woman who mopped floors.

The ridiculous nature of that should have made her laugh.

Instead, it made something tighten behind her ribs.

She knew men.

She knew what they wanted when they were scared. She knew how quickly admiration became appetite in their eyes. She knew the soft voice men used when they wanted permission to become worse than they planned. She knew the smell of rooms where choice was removed and replaced with something far more dangerous.

That knowledge lived in her body like old shrapnel.

So the feeling that moved through her now was not welcome.

It arrived as a crack in her concrete where something living made the poor decision to grow.

Vira looked down at her hands on the scarred plastic table. Cleaner's hands. Small cuts around the nail beds. A faint chemical dryness she could never fully wash away. Hands men once treated as available because everything else about her was made available by force, paperwork, hunger, threat, and the ugly mathematics of survival.

For years, feeling was a liability. Feeling gave men leverage. Feeling opened doors that should stay locked. Feeling made the body remember what the mind buried because soulful death was efficient.

But Gabriel Lerner did something deeply inconvenient. He became measured instead of loud.

She typed before she could punish herself out of it.

VIRUS_V: *you learn fast mr. brownstone.*

The instant she hit send, heat climbed into her face.

"Idiot," she whispered.

The first dryer buzzed.

She stood, folded her laundry, and placed it into her bag with deliberate care.

Not because the laundry mattered.

Because care mattered.

Care was how she controlled trembling.

In the brownstone on East 87th Street, Gabriel watched the reply arrive.

no dock

He read the rest slowly.

gray coat stupid. wear darker shoes

For the first time in hours, he laughed.

Not loudly.

Not happily.

But the sound had blood in it.

Arthur looked up from the secondary terminal. "What?"

"The ghost has operational standards."

Kael snorted through the encrypted audio. "You're smiling because a janitor insulted your wardrobe?"

"No," Gabriel said. "She changed the meet."

Arthur's expression sharpened.

Gabriel tapped the screen. "She understood what I was asking, then moved the terrain entirely."

"Meaning?"

"Meaning she's smarter than our collective panic."

Outside the brownstone, East 87th Street continued its moneyed tranquility performance. Tree-lined silence. Museum and symphony donors behind restored facades. The old city in its expensive coat, pretending history was something that happened at other addresses.

Two thousand, five hundred and seventy-two miles away, under the fluorescent wash of a gas station on El Camino Real, the real war was beginning beside windshield fluid, scratch tickets, bad coffee, and men too tired to look up from their phones.

Gabriel reread the final line.

you learn fast, mr. brownstone

Something inside him steadied.

Not trust. Recognition.

Kael's keyboard clattered through the channel. "Decoy script is scaffolding now. I can spoof your late-night search pattern in fifteen. Behavioral texture in forty."

"Make it ugly," Gabriel said.

"It will be gorgeous."

"Arthur?"

"I'm already pulling Becker," Arthur said. "Not public records. Patterns. Charitable giving. Conference travel. Committee resignations. Anywhere guilt leaves administrative dust."

"Maddy?"

"I'm thinking tactically about guilt," she said. "Which I resent."

"Good."

"No," she said. "Not good. Necessary."

Gabriel accepted that.

"Kael," he said, "I need the fake environment to keep querying Ohio and the tax hike. Let Farouk believe we're still obsessed with the thing she handed us."

"I'll make the fake you very committed to being wrong."

"That should be easy," Arthur said.

For half a second, the channel almost became human as everyone laughed.

Then the photograph pulled them back.

The manila folder.

The black notebook.

The red mark beside Gabriel's name.

Gabriel stood and went to the closet. The gray coat hung there. He left it where it was.

He chose a dark gabardine long coat instead.

His shoes. Polished. Italian. Expensive enough to be ridiculous in multiple languages. He stared at them, challenging them to fit the upcoming scenario, then changed into a pair of black walking shoes he used on the rare occasions he remembered that he needed more than the occasional stroll around the block.

Kael was still typing.

Arthur was still searching.

Maddy was silent. She was praying without admitting it.

Gabriel reached into the desk drawer and removed an envelope of cash, a burner phone, and a folded paper map of the Peninsula he hadn't looked at in seven years.

The absurdity of the map comforted him. Paper did not optimize. Paper did not predict. Paper waited.

He slept knowing that tomorrow would be one of the more interesting meetings he would ever take.

At 7:38 AM, Gabriel left the brownstone by the service entrance and walked east beneath trees that watched monied people do both glorious and shameful things. He flagged an anonymous taxi and told the driver to head to La Guardia.

Four time zones and fourteen hours later, Gabriel was closing in on a meeting that was critical to everyone's sanity.

At 11:07 PM, Vira badged through Building Three with a mop bucket, a gray utility cart, and no expression at all.

The night security guard looked up from his phone.

"Late," he said.

"Floor spill in analytics," she replied.

"Again?"

"Men with big degrees cannot hold coffee."

The guard laughed and waved her through.

This was what systems never understood. Not invisibility. Invisibility was too mystical a word. Vira was not invisible. People saw her constantly.

They saw a uniform. Function. Mop. Cart. Accent. Woman. Immigrant. Labor. They saw everything except the person arranging those attributes into a weapon.

She pushed the cart down the service corridor.

Building Three hummed around her. Air handlers and refrigerator compressors humming. Server closets behind locked doors emitting their own frequency. It was the living breaths of machinery.

At 11:12, she entered the supply room beside Receiving and turned off the overhead light.

Second silence began when the day shift left. The night crew split up by building and floor as security entered its lazy orbit between caffeine and boredom and candy crush. Three minutes when no one expected anything to happen because nothing ever did.

Nothing was a system too.

She wheeled the cart to the dock, propped open the service door for exactly nine seconds, and stepped out carrying a trash bag that was not trash.

The camera saw what it expected.

A janitor removing waste.

Systems loved expectation. They mistook it for truth.

She crossed the service lane, passed behind the delivery bay, and disappeared through the dark seam between two parked maintenance vans.

Vira slipped out of her coverall uniform and stashed it atop a tire of one of the vans. Underneath she wore a black thermal top, dark jeans, and darker sneakers.

At the far fence, she lifted the loose section of chain link she discovered three months earlier and never reported because reporting things was what people did when they believed institutions rewarded accuracy. She would also have no reason for knowing about that breach and she always had a reason for everything she did.

She knew there were no cameras looking in her direction.

Outside the campus, the night opened.

She walked south without hurry, though urgency was definitely surging within.

El Camino Real appeared ahead as a long corridor of commercial fluorescence and headlights. The old road. The King's Highway, buried under asphalt, franchise logo signs, tire shops, taquerias, pharmacies, tech shuttles, delivery vans, and people passing through each other's lives without contact or recognition.

California's great historical artery was now a stripe of errands.

At 11:23, she reached the Shell station at 39th.

The place glowed under hard white lights. Gas pumps. Scratched bollards. A rack of propane tanks and motor oil. A window display of energy drinks, jerky, glowing lottery standings, phone cables, and sunglasses nobody should buy sober.

A Tesla charged near the edge of the lot beside an old dented Kia with a garbage bag taped over one rear window. A man in scrubs bought coffee. Two teenagers debated over vape cartridges. A delivery driver leaned against his van scrolling through his phone with the postural collapse of a civilization that invented too many apps and not enough mercy.

Nobody looked at anyone for more than half a second.

That was anonymous safety.

Vira entered the store and bought a bottle of water she did not want. The cashier forgot her before she reached the door.

She stepped back outside and went to the side of the convenience building.

At 12:27, a rented Audi pulled in. A man in a dark coat got out and crossed from the far side of the lot.

Not gray. Darker shoes. He had listened again. Damn him.

Gabriel did not approach directly. He went inside first. Bought coffee. Black. Burnt. Undrinkable. Paid cash. Came out with the paper cup in one hand and nothing revealing in his posture.

He walked nonchalantly over to the side of the building.

He stood two yards away, looking toward the traffic.

Neither of them immediately spoke.

The night made ugly music around them. Pumps clicking. Compressors humming. A bus sighing at the curb. Tires hissing on El Camino. Somewhere, a man laughed too loudly into a phone and then stopped as if remembering laughter cost money.

"You're Virus," Gabriel said.

"You're brownstone."

His mouth almost moved into a smile.

Almost.

"We have five minutes," Vira said.

"Then I'll try not to waste them."

She turned her head slightly and looked at him.

Men said things like that all the time. It cost nothing to sound respectful when danger made humility fashionable.

But his voice did not reach for charm.

He was listening.

She reached into her jacket and removed a folded sheet of paper.

Gabriel did not take it immediately.

"Becker moved his notebook," she said. "Not in his office now. He carries it. But copies exist."

"Where?"

His face changed before she said anything else. He understood his mistake the instant it left his mouth.

"Wrong question," he said.

"Yes."

"What do you need?"

That was better.

Vira felt the crack widen and resented him for it.

"I need your mirror," she said.

"My what?"

"What they show you. Fake system. Fake Ohio. Fake ghost. All of it. You give me map of lie."

"So you can compare it with what you see physically."

"Yes. Digital lie and physical lie. Between them, truth."

Gabriel nodded once in recognition.

Vira looked away first.

"Do not do that," she said.

"Do what?"

"Look like you just discovered a person."

The sentence struck him cleanly.

He lowered his eyes.

"I'm sorry."

She believed him.

That was inconvenient. Again.

A sedan pulled in too quickly to pump three. Music thumped when the door opened, then cut off. A woman in a blazer got out, irritated, phone pressed to her ear. The world continued being stupid and brash around them.

Gabriel kept his voice low. "Can you access the local archive?"

"No."

"Can someone else?"

"Maybe."

He waited this time.

She noticed.

"Building Seven keeps scans of compliance packets on local archive," Vira said. "Not main system. Old habit. Men make secret and then backup secret because they are proud of secret."

Gabriel almost smiled again, then wisely did not.

"I need one path," she continued. "Not into archive. Around it. Who asks for report. Who approves access. Who receives copy. Systems have doors. People have habits."

"I can find request patterns," Gabriel said. "Not from the real system. But from the mirror, maybe the absences. If they built the simulation fast, they may have copied the surface and missed the administrative cracks."

"Cracks are useful."

"Yes."

She handed him the folded paper and a USB stick.

This time his fingers touched hers.

She noticed.

She hated that she noticed.

The paper contained six names, two floor numbers, one recurring maintenance window, and a hand-drawn map of a corridor no visitor would ever see.

Gabriel looked at it once, then folded it into his coat.

"Do you have backup?" he asked.

"Do you?"

"Not enough."

"Then you understand."

The ridiculous thing was that he did.

"And the stick?"

"Becker notebook details and ramping spikes with structured matrix." She answered.

A man who lived in an $18 million brownstone in Manhattan standing beside a convenience store gas station on El Camino Real understood something about exposure. Not the same kind. Not the old kind. Not hers.

But enough.

Vira felt the warmth again. Small and unwelcome.

"Will I see you again?" Gabriel asked.

There it was.

Not manipulation.

Still, the question entered her body wrong.

Rooms. Men. Doors. Demanding. Desiring. Wanting. Needing. Using.

Her throat tightened, and for half a second, she was not in San Mateo but somewhere older, smaller, plain and airless.

Gabriel saw the change and took one step back.

The crack became pain.

"Yes," she said before she could stop herself. "If you do not be stupid."

"I'll try."

"Try harder. You are expensive to move."

This time he did smile.

Then he turned and walked toward the far edge of the lot to his rented sedan. Good. He did not look over his shoulder.

Vira stood at the corner of the building, bottle of water unopened in her hand, listening to the ugly hum of El Camino Real and the distant electrical pulse of the valley that taught the world to confuse capture with connection.

Her palm pressed once against the wall.

Not for balance.

For proof.

The body could still feel something besides threat.

That was dangerous. It was unacceptable. She was alive.

From somewhere beneath her, too deep to hear and too deep to stop, the first fracture moved.

Seismic activity, Vira thought.

Small at first.

Then structural.

She opened the bottle of water, took a long series of sips, and walked back toward the dark.

NADIA'S SATISFACTION

The fourth floor of Building Seven was silent save for the hum of the climate control system. The glow of Nadia Farouk's triple-monitor array lit her face. On the center screen, a line graph for Subject G-87 was doing exactly what she designed him to do.

The curve was jagged and full of the anxiety markers that signaled cognitive collapse and emotional engagement. High-latency logins. Erratic cursor movements. Three distinct drafts of a fake manifesto that Kael's script let slip into the buffer. Kael laughed as he loaded them.

Nadia leaned back in her Aeron chair, a cup of steaming green tea on her desk. She didn't smile. Smiling was for people who were surprised by success. She simply felt the warmth of a proof of concept approaching terminal velocity.

And yet, something bothered her.

The subject breakdown was textbook.

A spike in late-night logins starting at 3:14 AM. Cursor tremors clustering in mathematically elegant bursts. Escalation pacing that aligned almost perfectly with the psychological model she constructed six months earlier.

Too perfectly.

She took in the moment, her mouse hovering over the raw packet header. Were the latency and reactive triggers too consistent?

Her eyes narrowed slightly at a cluster of erratic mouse movements. For a brief instant, instinct moved ahead of ego. She considered initiating a deep-packet inspection.

Then the GPS tracker pulsed red.

Gabriel was moving.

A small neurological reward arrived instantly. Attention displaced suspicion with the clean efficiency of conditioned reinforcement. The prey was in motion. The loop was holding. She felt it all.

Nadia dismissed the alert window.

"He's so predictable. That's why I chose him."

PRIZMATIK SYSTEM LOG — 22:17:33.4

SUBJECT: FAROUK, NADIA
MODULE: HANA-ENHANCEMENT-B
ACTION: PARAMETER ADJUSTMENT — CONFLICT-RESULTANT THRESHOLD
RESULT: ENGAGEMENT UPLIFT +18.4%
NEURAL FEEDBACK INDICATORS:
– DOPAMINE RESPONSE: ELEVATED
– CORTICAL FOCUS: STABILIZED
– DECISION LATENCY: REDUCED
– REWARD ASSOCIATION: REINFORCED
BEHAVIORAL PATTERN: SUBJECT DEMONSTRATES INCREASED EFFICIENCY UNDER ESCALATION CONDITIONS

The internal tracking module from the Monitor & Isolate protocol flashed another notification.

G-87 broadcasted a breakdown event.

He was heading toward campus wearing a gray coat, seeking physical convergence with a hostile abstraction.

Nadia picked up her YeaLink phone and punched the speed-dial button for the Chief Psychology Officer.

"Joel," she said when he answered. "Tell me you're watching the G-87 feed."

A pause.

Becker's voice arrived filtered through exhaustion. Not the exhaustion of overwork. Something slower. Moral friction wearing grooves into a man who still understood what those grooves meant.

"I'm watching it," he said. "And this is exactly the threshold event I warned you and Marcus about."

Nadia's eyes stayed on the moving red dot. "You'll need to be more specific."

"You've collapsed the distinction between fixation and satisfaction."

She leaned back slightly.

"No," she said calmly. "I've discovered they're the same thing at scale."

"Don't do that."

"Do what?"

"Flatten pathology into engagement because the numbers rise afterward."

Nadia tapped the screen once.

"Joel... he crossed four time zones to complete the reinforcement loop physically. The platform is no longer something he uses intermittently. It has become psychologically central to his identity structure. That's not failure. That's total integration."

"That's dependency."

"At scale, dependency is indistinguishable from satisfaction. The nervous system doesn't care why it returns. Only that it does."

Becker exhaled slowly through his nose.

She could feel the old version of him, the boss within him, assembling itself on the other end of the line. The bureaucrat. The public ethicist. The man who once believed language itself could slow institutions down if used with tact.

"You built a system," he said quietly, "that only recognizes distress as successful engagement."

Nadia said nothing.

"Look at the biometric cascade," Becker continued, voice sharpening now. "Elevated stress hormones. Compulsive return behavior. Sleep disruption. Escalatory ideation. Five years ago, you would have classified this as addiction architecture."

"Five years ago," Nadia replied, "we didn't have enough longitudinal data to understand recursive attachment formation."

"He's coming to campus because he believes he's at war." Becker said loudly.

"He's coming to campus because the system made itself indispensable."

"That is not the defense you think it is."

For the first time, irritation crossed Nadia's face.

"It's not a defense, Joel. It's an outcome."

"He's unstable."

"He's engaged."

"He's enraged."

"And he is still returning."

The silence that followed felt old and exhausted.

Becker spoke again, but the force shifted slightly now. Not gone. Just worn thinner.

"What exactly are you planning?"

Nadia's eyes tracked the pulsing GPS marker.

"I already notified security to let him approach the loading dock."

Becker went still.

"Nadia..."

"I want the biometric readings when expectation collapses in real time."

"You're staging psychological humiliation as a live behavioral experiment."

"I'm observing a completed loop."

"He's a human being."

"He's a participant."

"No," Becker said, sharper now. *"He's a citizen. There's a difference."*

Nadia ignored that.

"He thinks he's an insurgent," she continued. "He thinks he's discovered a path inside the architecture. But he's just fulfilling the exact escalation sequence the system predicted."

"And if prediction itself becomes the destabilizing force?"

That gave her pause.

Tiny. Brief. Real.

Then she dismissed it.

"You still think systems become dangerous accidentally," she said softly.

"And you're beginning to enjoy proving they don't."

The line went quiet.

That one landed.

Nadia turned slightly in her chair, staring through the glass wall of Building Seven toward the dark campus beyond it.

When she spoke again, her voice was cooled.

"Come down to Seven, Joel. Watch the feed in real time before you condemn it. You spent years writing papers about theoretical engagement recursion. I'm showing you the first clean proof of concept."

"I wrote about containment models."

"And now you work for implementation."

The silence stretched longer this time.

Not surrender.

But fatigue.

When Becker finally spoke again, the fight shifted terrain.

"If security sees signs of escalation beyond verbal confrontation, they intervene immediately."

Nadia recognized the concession for what it was.

Not approval. Accommodation.

A man reducing damage because he no longer believed he could stop the machine.

"Of course," she said.

Becker did not answer immediately.

When he finally did, his voice sounded older.

"I'll be down in a minute."

The line disconnected.

Nadia looked back at the center monitor. The Satisfaction Index continued climbing.

Gabriel Lerner was approaching Menlo Park in a dark coat, moving toward a confrontation he believed mattered. Somewhere inside the architecture, a simulated version of his panic continued unfolding beautifully across mirrored systems and predictive pathways.

For one brief moment, Nadia considered reopening the packet inspection.

The behavioral symmetry still bothered her.

Then the GPS marker moved again.

The feeling passed.

She picked up her tea and took another sip. It was cool now.

She barely noticed.

The data was perfect.

ENCOUNTERS

The fourth floor of Building Seven remained lit long after most of the campus had dissolved into the stillness of wealth, code, and climate control.

Nadia Farouk sat alone before the glow of her monitor array, watching Subject G-87 stall three miles short of campus.

The red GPS marker stopped at a gas station along El Camino Real.

For twelve seconds she said nothing, then she leaned back slowly in her chair.

Interesting.

Not because the stop itself mattered. Subjects in escalation states frequently interrupted momentum once physical convergence became imminent. The nervous system recoiled before impact. The body understood consequences before ideology did.

What interested her was the timing.

The biometric cascade accelerated exactly as projected during the drive south. Elevated cortisol. Respiratory irregularity. Obsessive return behavior. The system tightened around him beautifully.

Then the marker stopped. And stayed stopped.

Nadia enlarged the telemetry feed.

The stress indicators plateaued. That was slightly unusual.

Most subjects entering collapse spirals either accelerated into confrontation or fragmented into erratic withdrawal behavior. G-87 instead appeared suspended. Still agitated and elevated but oddly stable.

Her fingers hovered over the command line for a packet inspection.

Again, the suspicious sensation of perfection in his collapse returned. Not wrongness exactly, but texture. Something about the behavioral symmetry felt overly resolved. The escalation curve was too satisfying. Nervous breakdowns were messier than this.

Then Gabriel's pulse rate spiked again, and the hesitation vanished beneath the warm neurological certainty of predictive confirmation.

"He's freezing," Nadia murmured.

The Shell station feed showed nothing remarkable. A dark rental sedan beneath fluorescent wash. Pumps. Traffic. Human exhaustion under California light pollution.

To Nadia, it looked textbook.

The subject reached the edge of confrontation and recoiled into paralysis.

A perfect 9.4.

She began typing.

```
SUBJECT G-87: PHYSICAL CONVERGENCE ATTEMPT ABORTED
PRIOR TO CAMPUS CONTACT.
SUBJECT DISPLAYING AFO POST-ESCALATION EXHAUSTION
RESPONSE WITH RETENTION OF PLATFORM FIXATION.
SATISFACTION INDEX HOLDING AT 9.4.
NO ADDITIONAL INTERVENTION REQUIRED.
```

She reread the entry, then closed the file.

Outside the glass walls of Building Seven, Menlo Park slept beneath eucalyptus fog and venture capital insulation. Thousands of systems continued humming through the valley while people convinced themselves they remained individuals despite the evidence to the contrary.

Nadia stood and stretched slowly. For one brief instant, she considered reopening the telemetry stream then dismissed the notion. Perfection always felt slightly artificial the first time it appeared.

Gabriel sat at an innocuous terminal bar in SFO watching planes taxi through the low coastal fog while the adrenaline slowly drained from his bloodstream. He quietly enjoyed the wet halo of the wing tip nav lights glowing red and green as the liquid air enveloped them.

He hadn't touched the Bloody Mary in front of him.

The USB stick and paper rested inside the inner pocket of his coat like a second pulse.

Around him, the airport performed its eclectic yet predictable theater. Luggage rolling behind harried phone-to-ear flyers. Business travelers pretended not to drink too much. People who'd missed connections slept across chairs while CNN murmured about instability somewhere else in the world from a hundred monitors pointed in every direction.

Gabriel barely registered any of it. He kept replaying the moment Vira stepped back from the wall beside the gas station and looked unflinchingly at him for the first time.

Not long, but enough.

Do not do that. You look like you just discovered a person.

The sentence lodged itself somewhere beneath his intellectual machinery and refused to leave.

He had spent months turning her into architecture.

The Ghost.

VIRUS_V.

An intrusion point in the resistance node. A systems anomaly. Useful. Interesting. Brilliant. But not fully real.

And then she stood beside him under the fluorescent wash of a Shell station holding a bottle of water while traffic hissed down El Camino

Real drowning out all eavesdroppers if anyone cared enough to pay them the slightest bit of attention. They hadn't even glanced in their direction.

She'd been tired yet alert. Funny in dangerous little flashes, all the while carrying enough embedded fear inside her body to make flinching look involuntary.

Gabriel closed his eyes briefly.

He saw her recalibrate when he asked if he would see her again. She watched him carefully after he stepped back.

The realization of her guarded nature unsettled him more than the thumb drive in his pocket.

Prizmatik spent years learning how to reduce human emotions and interactions into predictive behavior, but the gas station revealed something the models still struggled to metabolize fully. Not data. Not interaction. Not sentiment mapping. Presence and bodies in space and time. The pause before someone speaks. The tiny involuntary recoil and the way silence itself changed texture between two people standing around a corner of a gas station. Body language.

The architecture still leaked through bodies and the way the slightest posture change and facial deviance relayed more information than remotely tracked biometrics ever could. That was Prizmatik's weakness.

Gabriel opened his eyes and finally took a sip of Bloody Mary. It was terrible. He smiled. Not because they were winning. Because for the first time in months, the world no longer felt entirely synthetic.

His burner phone buzzed once.

Kael.

YOU MISSED THREE PERFECTLY GOOD CHANCES TO LOOK PARANOID ON CAMERA BTW

Gabriel stared at the message.

Then typed back:

SHE TOLD ME MY SHOES WERE STUPID

The response came instantly.

OMG YOU ARE IN LOVE HA!

Gabriel laughed despite himself and slipped the phone back into his pocket.

It wasn't love but something much more intense. Recognition, respect and admiration.

At 7:42 AM, Vira sat cross-legged on the narrow edge of her bed while dawn slowly entered her apartment through the cheap blinds, most of the slats bent, which sent the early-morning light into pale, irregular, fractured stripes across her apartment.

She showered twice. The smell of industrial cleaner lived in her skin no matter how hard she scrubbed. There was something else she tried to scrub away. That wouldn't go down the drain either.

The apartment was small enough that she could almost touch everything important. Sink. Hot plate. Laundry basket. Secondhand bookshelf. Window. Locks.

Especially the locks.

Every few minutes her eyes would wander to the locks without realizing she was doing it.

Outside, Redwood City was in full force of its daily transformation towards productivity. Delivery and garbage trucks rumbled by. Early commuters in cars, buses and CalTrans sticking to schedules. Coffee grinders whirring in apartments and kitchens. Espresso machines hissing in drive-through stands. Tech shuttles inhaling human beings before carrying them south toward systems large enough to consume their names.

Vira held a chipped ceramic mug between both hands. The tea inside tepid from being ignored. She could not sleep. That truly annoyed her. She disliked when her body behaved independently from discipline. Exhaustion made people sentimental. Sentimental people made mistakes. Mistakes opened doors that should remain closed forever.

Still...

She kept replaying small things. The way Gabriel listened without interruption. The way he corrected himself after asking the wrong question. The way he had stepped backward immediately when fear crossed her face. Most men stepped closer.

That was what her body remembered first.

Pressure. Advance. Rooms becoming small as tunnel vision wrapped her world too tightly.

Gabriel Lerner had stepped back. Respectfully. Instantly. Without the bruised male ego glowering. The memory moved through her in uninvited, ominously visceral ways.

Vira stood abruptly and crossed to the sink.

Досить!

She rinsed the mug though it was already clean.

Outside, morning widened slowly across the buildings.

You're Brownstone.

The nickname started as mockery abutting disdain. Now? She was no longer certain. That irritated her more than anything else.

Because somewhere between the gas pumps and the fluorescent humming and the smell of gasoline and overheated brakes, she experienced something she no longer trusted herself to live with safely.

Not operational, but personal curiosity: What would a man like Gabriel Lerner look like when he was not frightened?

That thought arrived quietly and remained. Vira stared at the sink. "No. Absolutely not!" she said to no one. That road led to dependency, compromise, softness, distraction. Every ugly thing systems used to enter people and rearrange them from the inside outward.

And yet... he listened to her carefully and with full comprehension. Not performed listening. Not wealthy-man curiosity masquerading as empathy. Not fascination with poverty worn like an expensive scarf. Actual listening.

The crack inside her widened slightly.

Vira closed her eyes as confusion entered her thoughts.

For years she survived by reducing herself into function. Cleaner. Invisible labor. Residue collector. Ghost.

Useful things endured.

Women who wanted things got left behind or hurt. She wouldn't allow either.

But somewhere beneath all the operational discipline and survival mathematics, another thought moved carefully through the dark:

What if he keeps stepping back when it matters?

The apartment became very still. Vira opened her eyes, angry with herself now.

"Idiot," she whispered, but the word lacked conviction.

On the east side of the 101, Prizmatik's servers continued humming beneath billions of interactions, predictions, compulsions, and engineered desires and irrational rage fueled by algorithms.

Meanwhile inside Building Seven, Nadia Farouk watched peacefully, with no remorse. She did not notice the microscopic inconsistencies spreading through the architecture. Not code but people and the disruptive way that they overturned systems simply because of their unpredictability and desire to take back their agency.

Residue always was dangerous.

SLCNMSTER ↻ 3

It began, as most important things in Herbert's life began. With a feeling.

Not an idea exactly. Not yet. More like the ionized electrical charge before a thunderstorm, when the atmosphere knows something's coming, but the barometer hasn't caught up. Herbert learned to trust that charge. The charge had never been wrong. The charge was, in Herbert's private cosmology, the difference between him and the credulous masses who scrolled their feeds like grazing animals, consuming whatever the algorithmic trough delivered without once asking who filled it and why.

He posted at 11:47 PM. A thread of forty-one paragraphs. It was his best work since the Prizmatik data harvesting analysis three weeks prior, possibly his best work ever, meticulously documented and sourced (seven were his own threads.) Herbert considered this point and set his concern aside as irrelevant to the essential truth of the argument, his dissection of what he was now calling:

THE PRIZMATIK EMOTIONAL AMPLIFICATION MATRIX: HOW THEY WEAPONIZE YOUR OUTRAGE AND SELL IT BACK TO YOU AT A MARKUP!

He removed two of three exclamation points from the title. He wasn't an extremist.

The engagement counter sat at zero. He watched it for several minutes more, blinking only when his eyes burned, lest he missed the counter tick over.

He reached for the Mountain Dew. It was three-quarters full and almost cold, which meant it was early yet, which meant there was time, which meant the thread simply hadn't found its audience. This was the nature of important work. Important work required patience. Important work required a willingness to sit in the dark with his convictions while the world slept.

Zero.

He ate four Oreos. Double Stuf. Then a package of Ho-Ho's.

Zero.

Then: *one.*

He did not sleep that first night.

This was not unusual. Sleep, for Herbert, was always a negotiation conducted under duress, his brain presenting an agenda of unresolved grievances that his body was obligated to sit through before anything resembling rest could be achieved. But that first night the negotiation didn't happen at all, because the engagement counter moved to one and then to seven and then, in the time it took him to refresh his Mountain Dew from the second two-liter he kept on the floor beside the gaming throne for operational continuity, to thirty-one.

By two in the morning, it was ninety-four.

Herbert sat very still in his gaming throne, which registered its objection to his stillness with a long, pressurized groan. He ignored it. He was reading the replies with focused attention. Most of them were the usual – dismissal, three accusations of mental illness, one person calling him a hero, and TRUTH_ARTILLERY_88 sharing the thread with commentary that added nothing but introduced it to new audiences. But there were others. Accounts he didn't recognize, people who found the thread from vectors he couldn't trace without effort, and who were engaging with it not as an object of ridicule but as a foundation. Building on it. Extending it. Treating his documented observations as the starting point for their own.

One reply stopped him. It wasn't louder than the others. Not more aggressive. It was the opposite. It was measured.

It was referenced to paragraph twelve. Not the tone… the actual claim. It didn't agree. It didn't dismiss. It asked a question that suggested the person on the other end had followed the thread all the way through.

Herbert read it twice. Then a third time.

Someone was actually reading.

He opened Lotus 1-2-3.

He labeled the new tab: AMPLIFICATION EVENT – PRIMARY DOCUMENTATION.

He entered the data with hands that were, he noticed distantly, trembling slightly and damp with sweat. Not from fear. From a physiological response that acknowledges that he has been right and the world is finally, belatedly, beginning to catch up.

He minimized Lotus 1-2-3.

He opened a new thread. He considered replying directly. He didn't… yet.

Upstairs, at some point in the early hours of the first morning, the television turned off. The silence that followed – the signature quality of quiet that meant his mother was going to bed – settled over the house like a change in air pressure. Herbert's shoulders performed their Pavlovian contraction. He waited. The third stair creaked its familiar register. The bathroom. The flush. The tap. The shuffle back. The bedroom door.

He waited thirty seconds then turned back to the screen. The counter was at two hundred and sixteen.

By noon of the first day he'd forgotten to eat breakfast in any formal sense, though the archaeological evidence suggested ongoing contact with a party-sized bag of Cool Ranch Doritos, the remaining Oreos, and three gas station Slim Jim's of uncertain vintage he discovered in the left pocket of the hoodie he'd been wearing since Tuesday, which was now, depending on how you counted, either yesterday or the day before yesterday, a distinction that stopped mattering sometime around four in the morning.

The thread now numbered a thousand and twelve engagements.

Herbert stood up from the gaming throne – which expressed its gratitude at the redistribution of weight with a sound like a small animal – and paced the length of the basement in the lava lamp's amber light. Four steps to the Gateway 2000 tower. Four steps back to the gaming throne. The carpet was compressed into a permanent track from years of this exact pacing, a groove worn into the fibers by the weight of Herbert's thinking.

He was being seen and, somehow, though improbable, understood.

Not by people who mattered. Not yet. Not by journalists who should have been covering this, not by the congressional offices he'd emailed innumerable times without response, not by the tech commentator with four hundred thousand followers whose recent Prizmatik analysis he found directly derivative on points he'd made in a thread eighteen months ago that had received, at its peak, forty-one engagements – but by someone. By someones. By a growing constituency of someones who were reading his words and finding in them, the shape of something they felt but could not name.

He sat back down and opened a new thread.

The second night arrived without announcement. Herbert hadn't noticed the transition from afternoon to evening in any conscious sense. The monitors and lava lamp's amber glow simply became the primary light source at some point, the narrow basement windows going dark that he barely registered peripherally. The monitors glowed. The counter climbed. The gaming throne groaned every time Herbert moved a fraction of his weight.

He was, by this point, operating on something that was between sleep and wakefulness but existed in the emotionally charged territory between them, a state he visited before, during his most intensive research periods and that he privately considered his peak cognitive environment. Every other noise fell away in this state. The din of ordinary consciousness – the awareness of his body, the awareness of the house and his mother above him, the awareness of time as a linear phenomenon with obligations attached – all of it receded, leaving only the screens and the thread and the gathering momentum of being right in the public eye for the first time in a very long time.

The Doritos bag was empty. He did not register that fact but reached for the flaming hot Cheetos anyway. The Mountain Dew was warm and flat. Again. He drank it anyway.

At some point during the second night he posted something that was, stripped of the phrases "globalist engagement architects" and "weaponized dopamine syndicate" and "sheeple will scroll past this but the *real* people know" – stripped of the seventeen tabs of supporting documentation of which four were reliable, three were his own prior threads, six were forums he himself posted in – stripped of all of that was a substantially accurate description of a 400 millisecond delay in Prizmatik's egress layer that was not a latency issue but a deliberate throttling mechanism designed to hold high-sentiment outrage packets until the platform's bio-sync refresh could amplify their impact.

He was, in the essential analysis of all of this, correct.

The thread received four hundred and seven engagements in the first hour.

It was somewhere during the second day that his mother came downstairs.

He heard her on the stairs – the third one, its familiar squeaking register – and performed the window-minimization with the speed of long expertise. Three windows. Two seconds. The naked women on his smaller monitor hidden. He had been minimizing windows for decades and developed a genuine proficiency in the motion.

She appeared at the bottom of the stairs in her housecoat. The one with the small blue flowers, faded from hundreds of washings, into something resembling gray. She was carrying a plate. Two pieces of toast, buttered, and a glass of orange juice, the kind from the cardboard carton in the refrigerator, pulp free, the kind he preferred since he was nine years old.

She stood at the bottom of the stairs and looked at him. At the monitors. At the archaeological layers of the recent sessions on the floor. The Doritos bags, cherry pie tin, Ho-Ho and Twinkie wrappers everywhere, the pizza boxes stacked in every available space, the Slim Jim wrappers, the Mountain Dew and Jolt casualties. Her face held an expression he could not read and did not try to.

She set the plate on the edge of the cracked door in the narrow margin of space between keyboards, with careful precision. She had been leaving things within reach of her son for a very long time.

She glanced at him once more. Then she went back upstairs.

The third stair creaked.

Herbert looked at the toast then back to the screens. The counter moved again. He turned back to the toast as it went cold.

By hour forty the thread had become something he could not have anticipated nor could he fully process. It developed its own ecosystem. Other accounts were building sub-threads off his core documentation, extending his arguments into territories he hadn't explored, some of them accurate, some of them alarmingly incorrect, some of them combining his insights with superstructures of elaboration that he found simultaneously flattering and epistemologically horrifying. He was being cited. He was being quoted. He was being misquoted in ways that made his actual arguments sound more unhinged than they were, which he found deeply ironic given that he specifically removed two exclamation points from the title.

He posted corrections. The corrections were amplified at the same rate as the original. He posted clarifications of the corrections. The clarifications were amplified.

The engagement counter stopped being a number he could hold in his head and instead, became a rolling, living thing, a tide that he was no longer directing but simply riding, his hands on the keyboard producing words that became posts that became engagements that became more words, the cycle feeding itself in a loop he could feel but could not see, the way you cannot see the shape of a wave when you are tumbling inside the wash.

He was, in that moment, the most engaged with the world he'd ever been. He was also completely alone in a basement at three in the morning on the second day, wearing a hoodie that had not been washed for a very long time, eating toast that had been cold for eleven hours.

He did not know these things were related.

At four in the morning of the second night, he posted his most accurate thread yet.

It was also his most unreadable. Forty-seven hours did something to his prose that he could not perceive — a loosening of connective tissue between ideas and a tendency toward capitalization that migrated from emphasis to frequency with a rhythm that pushed an urgent, percussive quality. He could hear the truth so clearly but forgotten that the person he was telling it to could not hear it at his volume.

The thread contained, in paragraph nineteen, a specific and substantially accurate description of a conflict-resultant feedback loop that a behavioral scientist and executive forty minutes up the peninsula had been documenting in a sandbox environment for six weeks.

It contained, in paragraph thirty-one, a description of a physical residue trail — discarded documents, unlocked monitors, the informational debris left behind by institutions that thought they were invisible to their own cleaning staff — that was so close to one woman's actual methodology that, if she'd read it, she might have felt the chill of being seen by someone who didn't know she existed.

It received eleven hundred and forty engagements in the first two hours. It was shared by TRUTH_ARTILLERY_88 with the commentary: *bro said everything fr.*

The top reply called Herbert a schizophrenic.

The second top reply called him a great patriot to the nation.

The third top reply came from an account that was created forty minutes after his thread went live, a shadow profile had no prior history, and would be purged within seventy-two hours; it said: *you're getting close. be careful.*

Herbert stared at the words at 4:23 AM. He stared at it for a long time. Energy surged through his tired limbs. He leaned towards the monitors and typed:

"Source?"

He was not alone. Not this time.

He leaned back into the gaming throne. It groaned, again. The lava lamp moved shadows of its globular shapes across the ceiling. The monitors glowed. The counter moved.

Herbert reached for the Mountain Dew. It was empty. He didn't remember drinking the rest of it.

He dropped the bottle onto the floor to wallow with its kin and turned back to the screen. He felt, in the electrically charged space behind his eyes, the conviction of a man who is certain – more certain than he had ever been about anything – that history was finally catching up with him.

He opened a new thread and began to type.

PRIZMATIK SYSTEM LOG: 04:23:17.9

USER: SLCNMSTR69
USER ID 27AX9715PLD667
CONTENT CATEGORY: CONSPIRACY/PLATFORM CRITIQUE/ALGORITHMIC AWARENESS
OUTRAGE COEFFICIENT: 9.8/10
ACCURACY INDEX: 71%
REACH PROJECTION: EXCEPTIONAL
ENGAGEMENT TRAJECTORY: ACCELERATING
EMOTIONAL ACTIVATION SCORE: 9.6/10
INTERVENTION: SOFT FRICTION APPLIED
MESSAGE VECTOR: PEER-LEVEL RESPONSE – SYNTHETIC PEER CONTRUCT ACTIVE
OBJECTIVE: MAINTAIN INVESTIGATION LOOP
RECOMMENDED ACTION: AMPLIFY
AMPLIFICATION MULTIPLIER: 580%
NOTE: ACCURACY INDEX CRITICAL – APPROACHING THRESHOLD
NOTE: USER UNAWARE OF AMPLIFICATION
NOTE: USER APPROACHING SIGNAL THRESHOLD
NOTE: DO NOT INTERRUPT MOMENTUM
NOTE: USER SLEEP DEPRIVATION INDEX: SEVERE
NOTE: USER RETENTION PROBABILITY – 99.7%
NOTE: HIGH-FREQUENCY USER NEARING BIOLOGICAL LIMITS
NOTE: CONTENT PERFORMANCE UNAFFECTED BY BIOLOGICAL STATUS
SYSTEM STATUS: WITHIN OPERATIONAL PARAMETERS

Upstairs, the house was quiet.

The third step creak waited patiently for morning.

REFLECTION LAYER

BUILDING SEVEN

The lobby of Building Seven was designed to be a cathedral of bespoke power.

Glass. Steel. Wood and marble. The three-story living wall rose through the center of the atrium. The vertical forest had been created by consultants who believed nature could be improved through sponsorship agreements.

During daylight hours it was populated by founders, investors, influencers, and executives. It echoed with the sound of $300 custom sneakers on polished concrete and the low conversations of people who believed they were optimizing the human experience even while the language between the words spoke of disruption, optimization, and scale.

At 3:12 AM it belonged to something else entirely.

The enormous chamber breathed softly through hidden ventilation systems. Standing alone beneath the living wall, Soo-Ah Kim - Suwa to twenty-two million followers - felt strangely small even though her reach was global.

Somewhere above her, an elevator chimed. Suwa looked up. No floor indicator changed and no doors opened. Then there was a thump down a different landing. The sounds echoed through the atrium... then disappeared.

She frowned.

A moment later one of the information displays near reception flickered, going black, then gray and white. A fuzzy profile photo appeared of a smiling woman she didn't recognize. The image vanished before she could focus on it.

The screen returned to the rotating corporate slideshow.

WELCOME TO TOMORROW

EMPOWERING HUMAN CONNECTION

BUILDING COMMUNITY THROUGH AUTHENTIC ENGAGEMENT

Suwa stared at the display for several seconds then looked away.

The building seemed unusually awake tonight, as if it itself was aware. That distinction bothered her more than she cared to admit.

The living wall stretched upward through the darkness. Thousands of leaves. Thousands of tiny green surfaces quietly converting light into oxygen.

One section near the eastern edge curled beneath a failing ultraviolet fixture. A couple of leaves were dying.

Nobody noticed… yet.

Soft uplink indicators blinked from recessed equipment closets covered by mocha-colored glass. Somewhere behind the walls, data moved in cascading rivers large enough to shape elections, ruin or foster relationships, sway markets, and incite riots.

She was Bisou's owner to every brand manager around the globe. A product so successful that Prizmatik had given her Creator-in-Residence status and a luxury sleep pod to inhabit, a distinction she spent six years cultivating. Now she wasn't entirely certain why.

Her charcoal-gray leggings carried her 미래 logo across one hip in reflective silver. Her matching sports bra featured a copper-threaded geometric pattern designed to imply neural optimization. The entire outfit was provided by a wellness company that paid seven figures annually for product placement. She appeared to be exactly the person her audience expected. The version of herself that actually existed felt considerably less impressive.

Her eyes burned. Every blink left a ghost image of scrolling feeds suspended behind her eyelids. Her skin felt stretched thin beneath expensive makeup and twelve consecutive hours of artificial lighting. She hadn't slept properly in weeks.

The Ohio Tax Hike thread remained open on her phone. She read it six times. The comments weren't important but the engagement curve was.

Something about it bothered her. The post should have exploded. Instead, it stalled like a sagged spinnaker desperate for wind. Every surge flattened. Every wave lost momentum. Every spike was followed by a sudden appearance of unrelated content flooding user feeds.

Dog videos. Cooking videos. Travel reels. Yoga clips. Anything except the original discussion.

At first she dismissed it as coincidence. Now she wasn't so sure.

Her Samsung vibrated.

That device never appeared on camera. It never appeared in sponsorship contracts. It wasn't associated with any public account that her followers knew about. She called it her criminal phone. It was the phone she used when she wanted answers instead of engagement.

A packet capture continued running in the background. Numbers streamed across the display. Latency. Routing behavior. Traffic prioritization.

Minor anomalies that bore no conclusive points. There was simply nothing she could prove.

An elevator chimed, the sound echoing again through the enormous lobby.

The doors opened. A woman in janitorial clothing stepped out.

Her heavy gray overalls and practical boots with hair pulled back with severe efficiency spoke volumes about exhaustion as she pushed her cart loaded with cleaning supplies before her. The woman ignored everything expensive.

She ignored the art, the architecture, and the giant neon FUTURE sign suspended above reception.

Instead, she studied the floor. The scuff marks. The dust. A coffee stain beside a seating area. The world revealed itself to her as a collection of imperfections waiting to be corrected.

Suwa found herself watching as the woman crossed to the living wall.

She saw the browned fern leaf, strolled over, reached out and plucked it. She folded it carefully. Once. Twice. Again. Until it became a tiny square then slipped it into her breast pocket with poised intention.

She finally looked up. Their eyes met briefly yet something about her expression unsettled Suwa. It wasn't admiration. It was recognition. It was the kind of look a mechanic gives a machine the moment they understand the problem.

The woman turned and continued towards a corridor at a practiced pace. The scent of industrial lemon cleaner following in her wake.

Suwa glanced around the pristine lobby. The artificial rainforest fragrance suddenly smelled ridiculous. The make-up slathered serenity. The inspirational messaging. The optimization language. Everything felt indirectly intentional. As if the entire building depended upon people never looking too closely at the manipulation being foisted upon them.

The janitor disappeared.

Suwa remained beneath the living wall for another minute thinking about dead leaves. She thought about Ohio and about elevators that chimed but didn't open. She was thinking about why a janitor seemed more grounded in reality than anyone she'd met on the executive floors.

She headed upstairs, the feeling of trepidation following like an elusive shadow. One of those shadows that as soon as you turn to look is gone.

The Level Four Creator Lounge overlooked the atrium through a continuous wall of glass. At this hour it belonged entirely to her. The lockdown event was still running. Millions of followers continued consuming the carefully managed illusion of spontaneous authenticity.

Her ring light glowed beside her. Her public phone streamed. Her private Samsung continued sniffing packets from the guest network.

Every upload generated results. A variable lag between 400 and 520 milliseconds.

The precision bothered her. Networks didn't behave like that. Congestion fluctuated. Traffic shifted. Hardware hiccupped.

But perfect hesitations? That implied intent. It implied evaluation. And it seemed to target individual accounts differently yet with consistency depending on the user, as though the system hiccupped to consider whether the user should be allowed to speak. Hers was lagging at 480 milliseconds.

Four hundred and eighty milliseconds. The number appeared again.

Suwa stared at the packet trace.

She uploaded another image. Then a video, then a song.

Four hundred and eighty milliseconds. Again. The same delay.

Not approximately nor close. Exactly.

She ran it a fourth time. Then a fifth. Then a sixth.

Four hundred and eighty milliseconds.

Every time.

A cold sensation moved through her stomach.

Networks did not behave this way. Congestion fluctuated. Hardware failed. Packet routes changed. Humans made mistakes.

Perfect consistency belonged to intentional systems. Or intelligent ones.

She enlarged the capture window. The delay appeared between receipt and distribution.

It wasn't in the transmission loop or the delivery algorithm.

It was decision.

Something inside the system was pausing before allowing the content to proceed. Reading it. Evaluating it. Judging it. Then permitting it to exist.

The realization left her staring at the screen longer than she intended, as though some primitive instinct suddenly recognized the shape of a predator hidden in tall grass.

The packet log updated.

480ms.

She ran one more test and uploaded a plain black square containing absolutely nothing.

The delay disappeared. Only 120 milliseconds.

A new line of text appeared beneath it then vanished in a fraction of a second before she could read it.

Suwa leaned forward. "What the hell was that?"

No answer came. Only the soft hum of Building Seven and the feeling that she just watched something notice her looking.

She did it again - the black square. The delay was back... 480 milliseconds.

The message flashed again but this time she was ready for it: `ANOMALY REVIEWED.`

Her skin crawled. She leaned back against the glass.

Below her, Building Seven glowed like a sleeping dragon.

The Ohio thread remained open. She stared at it. The engagement suppression. The strange timing. The identical per-user delays. The feeling that conversations weren't being moderated. They were being steered and redirected. It was a subtler form of control and far more effective.

A soft mechanical whir drifted down the corridor. Suwa looked up immediately. The sound grew louder. Steady. Rhythmic, as it approached her position.

Her hand moved to dim the ring light. The corridor darkened.

In the reflection of the glass, she saw a figure emerge. It was the janitor from earlier. Pushing a large, motorized floor buffer. The machine hummed across the floor, bringing the surface to an otherworldly level of polish. The woman moved with detached concentration, yet there was an underlying tension within her movements that said more. She was cleaning evidence of human presence, removing traces and correcting imperfections.

She stopped beside a small circular table. A table where Suwa had left a spare battery pack and an extra gimbal. The woman ignored the equipment. Instead, she reached into her pocket and placed something on the table. Then looked directly toward the shadowed corner where Suwa stood.

This time there was no ambiguity. She knew. The object sat beneath the lounge lighting.

Pink. Rubber. A chew bone. Suwa's pulse stumbled. She knew the toy instantly. Three months earlier, she reviewed that same product during a Bisou segment. The engagement numbers had been extraordinary. The campaign became legendary inside Prizmatik as the textbook example of joy activation.

The woman tapped the bone twice while looking into the darkened corner. Then she turned and resumed buffing the floor.

By the time Suwa reached the hallway, the janitor was gone. Again.

The floor buffer sound vanished into the distant maze of corridors.

The dog bone remained. Waiting. A strip of thermal paper was taped beneath it. Suwa peeled it free. Read. Then read it again.

```
USER_ID: @SUWA_TECH_OFFICIAL
PROTOCOL: AFFINITY_MASK_77
STATUS: RE-EVALUATING
FUNCTION: DOG ASSET ACTIVE
FUNCTION: DEPLOY ON ESCALATION
```

The room tilted as her stomach seized. Memories flooded forward. Brand calls. Executive congratulations. Marketing awards. Campaign recaps. Celebratory dinners.

The dog. It had always been the dog. Whenever engagement exploded. Whenever metrics exceeded projections. Whenever executives praised her instincts. The fucking dog.

The horrifying truth emerged that she was being used. Every major engagement event. Every viral spike. Every celebrated campaign.

What if they weren't amplifying her content? What if they were deploying it? What if she wasn't driving attention? What if she was redirecting it?

The realization hit with almost physical force. She wasn't a creator. She wasn't even the product. She was packaging. Noise-canceling foam wrapped around inconvenient truths. A smiling distraction with a pink dog, delivered at the very moment people started asking questions and the machine required silence.

Her hands trembled. The thermal paper crackled between her fingers.

She looked toward the dark glass overlooking the atrium. Far below, the living wall remained illuminated. Thousands of healthy leaves. One dead one, quietly removed before anyone noticed.

The metaphor landed cleanly.

"You think I'm a curtain?" The words escaped in Korean. Barely audible. The voice didn't sound like hers.

She sank onto the floor as the smell of polishing fluid lingered in the air. For several minutes she simply sat there. Breathing. Thinking. Reassembling her understanding of reality as a decision sunk in.

She opened her Orchestra. Fourteen ghost accounts appeared, cascading windows launched simultaneously. The same accounts she built to manipulate visibility. The same tools she used to manufacture herself. The irony almost made her laugh.

Every lesson the system taught her. Every trick. Every optimization strategy. Every exploit. Now belonged to her.

She selected all fourteen accounts. Attached the thermal image. Attached the protocol receipt. Attached the packet logs. Attached everything.

A breadcrumb hidden in the network traffic caught her attention.

ZABAR_SYNC
Associated User: G-87

Another anomaly. Another ghost. Another witness. She included it then pressed <send>.

Nothing happened. Then the wall monitor activated. System diagnostics flooded the screen.

PRIZMATIK SYSTEM LOG: ANOMALY

SYSTEM ALERT: UNEXPECTED ENGAGEMENT ANOMALY
USER: SUWA_TECH_OFFICIAL
CONTENT TYPE: CRITICAL/LEAK
REACH MULTIPLIER: 1,400% [↑]
NOTE: SUBJECT_ASSET INVERSION DETECTED
NOTE: HEURISTIC DRIFT DETECTED
NOTE: FEEDBACK LOOP COMPROMISED
NOTE: ASSET NO LONGER PREDICTIVE
NOTE: SUBJECT WILL CONTINUE
RECOMMENDATION: CONTAIN

Suwa stared at the monitor. The recommendation appeared first. Not an investigation. Not a review. Not a verification but *suppress.*

The machine was deciding how quickly she could be erased.

Outside the glass, Building Seven continued humming through the darkness. Inside the system, the posts continued replicating, rapidly gaining global exposure.

For the first time in six years, Soo-Ah Kim wasn't manipulating her followers, generating likes and comments or using the algorithm. She was forcing it to react by exposing exactly what it was doing to everyone on the platform.

She thought about the leaf, quietly removed before anyone noticed.

Her Samsung vibrated. A notification appeared and vanished before she could screen capture it.

REVIEW COMPLETE

The screen went black.

FALLOUT

The silence in the executive boardroom wasn't the usual polished, corporate quiet. It was filled by a dense atmosphere. The kind that precedes a storm.

Zhen Sheng had seen the direction years ago. Not the specifics… not the architecture as it existed now… but the vector. The quiet shift from engagement to manipulation. But that wasn't enough.

At Prizmatik's scale, suspicion without evidence collapsed under legal language. Instinct didn't survive boardroom scrutiny.

He hadn't acted then but he would act now.

He didn't sit. He stood at the head of the table, the same mahogany surface where he toasted to the reunion of veterans and lost fathers, flown in for mutual benefit. He hadn't brought a tablet or a slide deck.

He brought only a documented multiplier-induced millisecond delay in the egress layer… the cocked hammer. It was in printout form… the entire code structure.

Marcus and David sat on either side of him, looking small.

"I didn't invest in a human emotion manipulation laboratory," Sheng began. His voice was low, but it vibrated with anger that made the glass walls hum. "I invested in a system designed to reveal connection, not foment anarchy and anger. I invested in the belief that if you showed people the truth and guided them to their own connections, the world would be better. I didn't lend you my capital and my name so you could build a machine that rewards emotional destabilization."

David started to speak, his hands moving toward a defensive gesture. "Zhen, the engagement metrics…"

"Do not speak to me of metrics!" The roar silenced the room. It wasn't the shout of frustration; it was the fury of a master craftsman seeing his work used as a manipulation weapon.

"I have spent over thirty years in security," Sheng continued, his eyes locked on Marcus. "I know how to identify a threat. I know how to model a catastrophe. And right now, the greatest threat to the stability of the networked world isn't a state-actor or a virus from the dark web. It is *Nadia Farouk.* It is the behavioral architecture she has built under your tenure and oversight. It is the conflict-resultant logic that values emotional storms over conversation. But… you already know all of this!"

Sheng leaned over the table, his knuckles pressing into the surface. The stoic, ribeye-searing, King Crimson-listening philosopher was gone. In his place was the Architect Emeritus of social media internet, and he was there to tear it down.

"You think you are untouchable because of the growth curves and popularity and global reach and revenue," Sheng said, his voice now a lethal whisper. "You think you can align your way out of the ethics with corporate justification. You are wrong. I hold the voting control of this company in case you've forgotten the bylaws agreement we all signed. And I am telling you now: The conflict inducing delays end tonight."

"Zhen, you can't just…" Marcus began.

"I can," Sheng cut him off. "If Nadia Farouk is not stripped of her authority across all assets and access of this company immediately… if her conflict-initiating modules are not purged from the production environment and the BMI feedback loops dismantled, I will call for a formal Vote of No Confidence in both of you."

David's face went pale. "The IPO... the investors... it would be a bloodbath."

"Let it bleed," Sheng said. "I would rather see Prizmatik burned to the ground than see it become the very pathology I spent my life trying to prevent. You have twelve hours. Bring her to bear, or I will dismantle this company."

Sheng turned and walked out. He didn't wait for a rebuttal. He didn't look back.

He took the elevator down, walked past the living walls that now felt like a funeral shroud, and got into his Maserati. He didn't drive north to The City and go back to Argus Secure. He drove around to Bedwell Bayfront Park. He sat beside the bay smelling salt, fog and primal earth, waiting to see if David and Marcus would choose to save Prizmatik or if it would die in the flames of its own hubris.

FRACTURE POINT

The founders' offices no longer felt like sanctuary. They felt like bunkers. Marcus stood by the floor-to-ceiling glass watching the Bay fog roll moodily over the Peninsula hills.

The first modules had been embarrassingly simple. Birthday reminders. Reunion suggestions. Military buddy matching. The anniversary engine that reminded people of first messages, first photos, first friendships. Millions of tiny prompts nudging people back toward one another. Most generated almost no revenue. Marcus defended every one of them. Connection wasn't a feature. It was the product. Then advertising arrived and the definition of success shifted.

David was already in the conference room, seated at the table at the far end position with only a notepad and pen.

He found himself staring through the glass wall overlooking the campus.

The sunlight glinted off Building Three and reflected on the surface of Taleb Pond below. Black swans drifted silently across the water.

Twenty-three years earlier, before investors, before quarterly guidance, before the company occupied enough real estate to require dedicated shuttle routes, he'd spent an entire weekend reading posts in a section of the Connection Library.

The name sounded naïve now.

It existed, not only because of the founders wish that the purity of what they built would have an archive of genuinely humanistic moments of connection to draw upon, but also because people kept asking the same question.

Is there anyone else out there going through this?

The answer, they discovered, was almost always yes.

One of the earliest sections, a group that formed organically, had been created by a woman in Nebraska undergoing treatment for ovarian cancer. David could no longer remember her real name. Only her screen name.

PrairieMoon.

She began by posting questions about nausea management, treatment schedules, and whether anyone else experienced a strange metallic taste after chemotherapy.

Three people responded. Then thirty. Then three hundred. Within months the group became something larger than anyone would have anticipated.

People shared simple recipes when chemotherapy and radiation destroyed appetites. They exchanged advice about insurance companies. They explained medical terminology. They sat awake through nights with strangers they would never meet posting back and forth simply to say - "hey - I'm here, and I see you!"

One husband posted updates every Tuesday after his wife's treatments.

A retired nurse from Arizona answered nearly every question posted.

Someone in Toronto organized meal deliveries for a member she had never met.

The entire thing operated without incentives. No rankings or followers or metrics.

Just people helping each other because they could.

David remembered reading through hundreds of posts late one night and realizing that the internet became something nobody predicted.

Community and connection.

Months later, the Connection Library became Groups.

Groups became Communities.

Communities became Affinity Clusters.

The names changed as the company grew.

The metrics arrived.

Then the dashboards.

Then the optimization.

The architecture remained but the purpose evolved.

David searched for her once years later out of curiosity.

He remembered typing the screen name into an internal search tool and discovering the account had gone inactive almost a decade earlier.

There hadn't been a final post or a goodbye and there was no announcement of her passing.

One day she simply stopped logging in.

Perhaps she survived. Perhaps she didn't. The platform never knew the difference.

The metrics recorded a decline in activity. A reduction in engagement. A dormant account.

The dashboards tracked the absence.

The system never understood the loss.

For years afterward he would occasionally think about her whenever someone described users as nodes, clusters, cohorts, demographics, or behavioral segments.

PrairieMoon had never been a segment.

She'd been a frightened woman in Valentine Nebraska who wanted to know if anyone else understood what she was going through.

And for a little while, thousands of people answered.

David looked down at the blank notepad in front of him, a knot in his throat that persisted after he took some deep breaths.

Somewhere inside the system, millions of people were still searching for the same thing they'd been searching for twenty-three years earlier. He knew this.

Is there anyone else out there going through this?

Marcus exhaled once and walked from his office, entered the room, and took the opposite end.

The door slid open. Nadia Farouk walked in first, her stride radiating kinetic energy. She believed she was about to be promoted. Behind her, Dr. Becker lagged. He entered the conference room looking like a man who knew what was coming.

"Marcus, David," Nadia said, her voice bright and earnest. "If this is about the G-87 breakdown, I have the full biometric overlay ready. The hesitation event at the Shell station is a textbook case of…"

"Sit down, Nadia," David said. He didn't look up.

His tone and volume were a cold bucket of water. Nadia froze, her hand halfway extended in greeting before slowly lowering it. She sat. Becker took the chair furthest from her, staring at the wood grain of the table.

"Zhen Sheng just left this office," Marcus said. "He didn't come to talk about hesitation events or joy amplification. He came to talk about 0x8F2A1C9E. The injection port. The fake tax hike in Ohio from six months ago. And the wreckage in the wake from all of that."

Nadia's eyes flickered with a microsecond of calculation. "Zhen sees threats because security is the only language he trusts! He sees backdoors and threats where there are only optimizations. The Ohio pilot was a necessary stress test of escalation dynamics. The platform stabilized after..."

David interrupted her and laughed. "Is that honestly what you tell yourself? Jesus Christ, Nadia, are you really so naïve... or is it arrogance? You've been inside your model far too long."

Marcus glared at her. "Zhen's calling for a Vote of No Confidence." For the first time, the empath of Prizmatik looked dangerous. "Against me. Against David. He wants your head on a platter, Nadia. He has demanded the BMI feedback loops dismantled by dawn, or he burns the IPO to the ground."

The silence that followed was absolute. Her carefully calibrated world was suddenly colliding with the reality of Zhen Sheng's wrath.

"He can't do that," she whispered. *"The growth metrics..."*

"The metrics are a lie Nadia! The Board believes the platform, as it stands today, is a liability" David snapped.

He nodded his head at Becker. "Joel. You've been quiet. You've been looking at the loop reports for months. Is Zhen right?"

Becker didn't look up. The clock on the wall ticked three times; mechanical strikes in the vacuum of the room that sounded louder than they were. Marcus shifted in his chair. The leather groaned softly, an ordinary sound in the middle of institutional collapse.

"Yes," Becker said, his voice was steady. "We aren't revealing light anymore. We're manufacturing dissent. Nadia isn't just optimizing engagement; she's engineering radicalization because it's the only way to keep the curve from flattening. We lied to the users. We lied to the Board... and we lied to ourselves."

Nadia's eyes grew cold with the sharp stillness of an academic predator who just found a fatal flaw in a rival's dissertation. "*You coward! You* signed off on every module. *You* took the equity. *You* sat in those alignment sessions and nodded while I did the hard work of saving this company!"

He thought in a flash of recognition that lasted a second but felt like a month: *There was a moment… one of a thousand small moments… when I knew exactly what I was looking at. The system flagged it. Ethics compliance failed it. I had the authority to stop it. I remember the cursor. Just… sitting there. Waiting for me to decide what kind of man I was. I told myself it was provisional. That I was creating space to study it. That I still controlled the outcome. I clicked approve.*

He rejoined the conversation. "Sheng doesn't need the money or the recognition. And he isn't silent anymore, or "only security" as you seem to believe. And all he has to do is say 'boo' and the entire party shuts down."

Marcus leaned onto the table and stared, unblinking, at Nadia. "He gave us twelve hours. Twelve hours to strip your authority and purge the conflict-initiation code from the production environment. I am advancing the timeline. You're done. Now."

Nadia stood and stared back at him. "If you do this? Engagement drops forty percent in twenty-four hours. The platform will look like a ghost town. You'll be admitting your prism is broken."

"It IS broken," Marcus replied. "Because of you. Because of me. Because we kept telling ourselves it was still the same thing."

David interrupted. "We're going to the Potemkin backup. Freeze the feed and tell the world we're doing a Core Values Audit. We take the hit now… or Zhen kills us tomorrow."

Nadia looked at the three men. She saw anxiety and determination in David, exhaustion in Becker, and desperate realization married to fury in Marcus.

"You think you can stop a virus by cutting off a limb?" Nadia said, her voice dropping. "I built this architecture to be self-sustaining. You purge me, the system crashes. I'm the only one who knows how to keep all those satisfied, engaged users from realizing they've been played."

"Then let it crash," a calm voice said from the doorway.

They all turned. Zhen Sheng stood in the threshold, Maserati keys in hand, his coat carrying the smell of fog.

"The emotional engagement and amplification delays end now," Sheng said, his eyes fixed on Nadia. "Pack your things. If you're seen on this campus after 4pm, I won't simply fire you. I'll hand the Argus Secure audit of your Conflict-Resultant logs to the Federal Trade Commission. Security is waiting in your office so you may collect your personal belongings."

She walked to her office with a rigid, icy composure, her mind already navigating the sub-layers of the architecture she'd spent years weaving.

The Dead-Man's Switch.

It was a beautiful piece of logic, a recursive conflict loop buried inside the low-priority affinity retention architecture tied to birthday notifications. If her specific administrative heartbeat wasn't detected for seventy-two hours, the system would automatically elevate every High-Arousal Negative Affect thread to the top of every feed globally. It would turn every minor grievance in every culture, in every language, into kindling.

"The company has taken possession of all assigned hardware, Ms. Farouk," the lead security officer said. He was a man Nadia willfully ignored for years. Another security officer waited outside her office. "Your internal and remote access credentials, corporate and private, have been revoked. You have five minutes to collect personal belongings. Then I'll escort you out."

Nadia felt a sudden, sharp chill. The staggering, ego-bruising realization hit home. Sheng hadn't just reacted to her. He anticipated her next moves. He'd mapped her emergency exit before she'd even built it. He hadn't just cut the limb. He sterilized her theater.

The drawers and slide tray were already open, her desk emptied of everything belonging to Prizmatik. The virus had been the physical breach, but Sheng was the systemic shut-off. There wasn't even a memory stick or paperclip in sight.

"I see," she said, her voice a thin, dangerous line.

An empty box sat on her chair. She gathered the few personal items on her desk.

As they escorted her toward the elevator, past the living walls that were now being digitally scrubbed of her influence, Nadia didn't think about the lost equity or the impending IPO. Residue and G-87 were, at that moment, the only things worth considering.

Gabriel Lerner was her only potential asset left. He was her subject who'd broken and proved her psychological manipulation tools worked exactly the way she envisioned. He was one who had come toward campus, desperate and fractured, hoping to weaponize the truth. If she could reach him, if she could feed his paranoia with the reality of what Sheng and Marcus were doing to the platform, she could turn him into a kinetic detonator.

"Enjoy the silence while it lasts, Marcus," she whispered as the elevator doors slid shut, sealing her out of the kingdom. "Static is louder than you think."

The flight back to JFK had been a blur of low-altitude clouds and high-tensile nerves. Gabriel sat in his brownstone, the Static sanctuary now humming with a different kind of energy.

The thumb drive from the Shell station sat plugged into the isolated, air-gapped terminal. It was scanned for incursion code and returned clean. That made him admire the Ghost even more.

On the 48-inch 6K display, the Ohio Pilot wasn't a graph anymore. It was a ledger of sins.

"They actually did it," Kael said over the mesh link, his voice sounding hollow. "They didn't nudge the tax hike story. They used a bot-net to simulate local outrage from accounts that didn't exist, then fed those outrage pings into the feeds of parents and community leaders. They engineered the physical violence, Gabriel. They weren't watching the fire. They built the room, soaked the walls with vitriol, and lit the match."

Gabriel said "It wasn't only them, Kael. Farouk designed the architecture, but Marcus and David let it breathe." He stopped. He was looking at a new notification on the screen.

A private, encrypted packet landed from VIRUS_V.

He opened the compressed file. High-resolution images flooded the screen as folder after folder opened automatically. Handwritten notes. Sideways scrawl. Ink on paper. More than a hundred pages.

Gabriel leaned closer, scanning the first page. He was looking at the entirety of Dr. Joel Becker's black book.

It wasn't code.

It wasn't alignment meditations or meeting notes.

It was a confessional in black ink on cream paper written in a notebook a continent away by a man who'd run out of places to hide.

LEGACY SYSTEMS

Zhen descended into a deeper layer into the repository tree. He was alone in his office at Argus, mining through older dormant repositories. It was dark. His staff had gone home hours before and the only thing left was the glowing array of monitors and the original code structures, built for growth and connection. He wasn't looking for bugs. He was reconstructing history, delving for something akin to legislative intent, to get back to the root of all things promising. Entire branches of Prizmatik's cognitive history still lingered there.

Dormant friendship recovery engines. Reunion networks. Legacy grief-support modules.

Abandoned systems whose names and functions no longer appeared on any operational or hierarchy or dependency schematic. They'd not been modified in years.

One source file caught his attention only because the comment block remained unusually intact.

```
// PURPOSE:
// Increase meaningful reconnection between dormant social
ties.
//
// SUCCESS METRIC:
// Conversations resumed after 90 days of inactivity.
```

He stared at it for a few ticks of the second hand of his Rolex, then returned to tracing the engagement pipeline. As he delved deeper into the past, the original-intent boomerang of good intentions circled back in comments fields.

Then...

```
// GOAL:
// Help users maintain meaningful long-term relationships.
```

Lastly, he came across the most meaningful comment fields he'd seen, which reminded him of why he had invested in Prizmatik in the first place:

```
// OBJECTIVE:
// Strengthen relationships through meaningful interaction.
```

```
// RESULT:
// Connection is the product.
```

He cycled through those innocuous notes, those developer comments, as he sat on a bench in the Southern California sun and realized that, in the years since his initial investment, the connection events he'd celebrated had become camouflage for a new ghostly agitation in the feed.

THE ARCHAEOLOGY OF GUILT

The 6K display was so bright it hurt Gabriel's retinas. He dimmed it until the scans of the black notebook looked like old parchment recovered from a library archive.

Gabriel's stomach performed a slow churn. He wasn't looking at a technical manual. He was looking at a ledger of cowardice and greed. His breathing was shallow and the air in the brownstone felt recycled and thin. He reached out, his thumb hovering over the glass of the monitor as if he could feel the indentation of pen on paper.

The first page was dated September 30th. Underneath it, Becker wrote:

Sept. 29th: During yesterday's meeting N called it frictionless radicalization. I called it a shit-storm blaze in a theater. She laughed and said theaters have exits. That's her simple answer to the mess she created. N thinks that users can simply walk away from emotional engagement.

Gabriel scrolled. The clarity and density hit him like a physical blow to the chest.

Becker mapped the variable delays as a cognitive gap. He found that if you delay a user's ability to post a rebuttal by just a fraction of a second, geared to their individual outrage patterns, their heart rate increased. Their High-Arousal Negative Affect spiked. By the time the post button became active, they weren't looking for a conversation anymore. They were looking to deliver a kill.

"He knew," Gabriel said aloud to no one, his words sounding like shattering glass in a cement-walled room. "The *bastard…* he knew exactly what those lags were doing to them."

Then, the tone of the book shifted. The visceral reality of Becker's internal collapse began to bleed through the ink. The handwriting grew jagged, the loops of the 'g's and 'y's dragging like tired feet.

November 22nd: I watched Subject 4412… a mother in Toledo, interact with a fake news injection. She spent six hours arguing with a bot Nadia designed to emulate her ex-husband. 4412's blood pressure readings (via her Prizm-watch sync) hit 160/100. I could have killed the thread. I could have adjusted the weight. I did nothing. I watched the engagement curve radically rise and

I thought about my stock options. I am a monster who's become fond of exceptional wine. I buried my guilt by bidding on 6 bottles of 99 Grange.

Gabriel felt a surge of nausea as he sat down at his desk; the custom server stack and thought about the $18M view from the window. He was part of the asset class that Becker's team was leveraging.

March 14th: N found 400ms lag wasn't enough. Some users adapted or rejected too quickly. She's now experimenting with variable latency bands tied to biometric escalation markers. She called it emotional elasticity mapping. I think she's teaching the system how to provoke users individually instead of by grouping or region.

The last ten pages of the scan were the most frantic. This was the extrication phase. Becker had begun mapping Nadia's administrative heartbeat and backdoor insurance policies.

June 10th: N built a Dead-Man's Switch. I need to find where she hid it. If she's fired, the system will vomit the worst of the Ohio Pilot concept globally in thousands of fake disasters and spin it out of control. I am tethered to a falling anvil, and it is my fault for not stopping N. I have to find someone, and I think I know who can expose N's bs without it landing on me. I need to find that woman. The cleaner. I believe she's the only one who doesn't have a stake in this. She knows where the exits are. I am sure of it.

Gabriel sat back. His chair didn't groan like Herbert's; it was silent, expensive, and suddenly felt like an overly tight cocoon. It was 4:12 AM.

He had no idea that in a glass office he'd never seen, Zhen Sheng was currently holding a knife to the throat of the company. He didn't know that Suwa was currently lighting a flamethrower on Menlo Park's most powerful human digital control center.

He only knew that he held Becker's confession in writing.

He reached for his phone, his fingers trembling so violently he nearly dropped it. He didn't call Arthur. He didn't call Kael. He opened the private link to VIRUS_V.

G-87: I read the black book. I see him, V_V. I see the man inside the machine trying to crawl out while the machine is spiraling down into free-fall. I feel bad for him because he felt a conscious emotional connection to humanistic reality until NF usurped him.

January 12th: Marcus asked me today if we could tune the empathy coefficient. He didn't use the word suppress but that was obviously the goal. He used 'optimize for resilience.' It's a corporate euphemism for turning off the part of the brain that keeps us from stabbing each other. I signed the memo and ordered the work. I greedily took the equity. I am a ghost. I am looking at the reflection of a man who has optimized himself into a farcical adaptation of moral control.

Gabriel read that and felt sick. In the Prizmatik architecture, a ghost was an account with no human rights, a data point used only to stress-test the breaking point of others.

He felt an ache in his jaw. He had been clenching his teeth so hard his face felt heavy. He reached out and shoved the monitor away, the heavy arm of the desk mount groaned. The motion was too fast, too violent; he knocked a glass of water over, the liquid seeping into the leather pad inset into his grandfather's desk. He didn't move to clean it. He just watched the water spread.

Gabriel heard his voice cut into his thoughts. "He knew! He knew they were turning the world into a cage match and he just... he just asked for a better view."

The final pages of the book were the most visceral. The handwriting wasn't just jagged now; it was frantic, the pen digging so deep into the paper it had nearly torn through.

March 14th: N found 400ms lag wasn't enough. She wants to introduce predictive conflict. Not only reacting to a fight but starting one before the user even knows they're angry. I can't stop her. The code is recursive now. If I pull the plug, the Dead-Man's Switch in the birthday announcement module triggers. It's a suicide vest made of algorithms. I am going to leave my book for the woman with the mop. If there is any justice left in this world, I believe she will know what to do with it.

Gabriel sat in the dark, the blue light of the terminal illuminating the damp spot on his desk. The clarity was complete. He had the weapon.

He stood too quickly as the room lagged half a beat behind him. He walked to the window. Outside, East 87th Street looked the same as it did yesterday. Quiet, expensive, and blissfully unaware that its collective nervous system was being rewritten by a tech influencer in the Mission District, a conspiracy theorist in a basement, and a security mogul in a Maserati.

He glanced at his phone. The notifications were beginning to stack. The orchestra Suwa launched was hitting the mainstream.

G-87: Vira. I've read it all. The Dead-Man's Switch is in the Birthday module. Becker is ready to flip, but he's terrified. If we hit them now, we hit them while the logic is unstable. Tell me where you are. We need to talk.

He waited.

The cursor blinked. A steady, mocking heartbeat.

Then, a reply scrolled into the terminal. It wasn't text. It was a 24-bit audio file.

Gabriel plugged in his headphones. He expected code, or a voice memo.

Instead, he heard the rhythmic, industrial *whir* of a floor buffer in a long, empty, echoing hallway. And then, a whisper. Low, steady, and full of the cold iron of Kharkiv:

"Come back, Gabriel. Bring book. We show NF what happens when furniture talks. I wait by curb at terminal."

THE CARD

Zhen looked at it, then looked at it again. He turned it over but there was nothing there. Just blank stock. A worn birthday card with a simple hand-written message inside that was deliberately left on his desk in the board suite atop his weekly packet.

Happy Birthday, Margaret.

Though thou art in heaven, we miss you still.

Remember to send reminders.

That was it and that was all.

He mused over its meaning for a while. Margaret meant nothing to him. Neither did the handwriting, yet someone had bypassed security into the executive suite and placed the card on his desk.

He set the card underneath the pile of documents awaiting his review, approval or rejection.

He saw the card again as he slid the last action memo towards himself. The realization that it was meant specifically for him landed uncomfortably.

He slipped it into his messenger bag.

PART III

MID-AIR STASIS

The cabin of the Gulfstream G650 was a pressurized tube of brushed aluminum, rich cherry wood and white leather. Nadia Farouk sat in seat 2A, her personal laptop used as her primary access point to the internet for the first time in eight years.

Below her, the American Midwest was a dark carpet of flyover states speckled by communities lit up by boulevards lined with streetlights and malls. She gazed out the window and didn't see geography; she saw an untapped laboratory of high-arousal negative affect - thousands of miles of subjects unaware that their emotional baseline had been the plaything of the woman currently sipping guava-scented sparkling water at forty thousand feet. To Nadia, the lights of Omaha and Fort Wayne weren't homes or lives; they were nodes in a predictive model. They were the low-engagement intellectuals and emotionally activated clusters she'd spent a decade manipulating.

She wondered, with detachment that bordered on angry euphoria, how many of the people in those houses were currently experiencing the AFO she pioneered. How many were staring at their screens, their heart rates climbing, their cortisol levels spiking, then the dopamine hit, as the cocked hammer of the egress layer held back their rebuttals, then released at the right moment? She felt a pang of phantom-limb syndrome for the data streams. It was an itch in her neocortex, a desperate need to see the real-time feedback loops of a million simultaneous indignities. Without the terminal, she was blind. Without the metrics, she was just a woman on a plane.

Her hands were still as they rested on the armrests, but her cuticles were raw. She'd picked at them until they bled during the taxi ride from Menlo Park to the private jet gate of SFO. She didn't feel the sting. Pain was just another physiological variable, and she had optimized to her personal threshold of discomfort long ago.

She was calculating as the plane neared its destination. Sheng thought he'd sterilized her theater, but he hadn't accounted for the recursive nature of her Dead-Man's Switch. The hidden Birthday Module code was a ticking clock. It was her masterpiece — a suicide vest made of algorithms hidden in the most innocuous corner of the UI.

By the time she landed at JFK, the Static would be loud enough to drown out even Sheng's Maserati. She imagined the headlines: a global spike in conflict-resultant interactions, a breakdown of the affinity mask, the glorious collapse of the Potemkin false front. She would walk into Gabriel Lerner's brownstone not as a vestige of hubris, but as the only person capable of turning off the noise.

The Gulfstream maintained forty-four thousand feet, the cabin pressure adjusting so subtly with the rise and fall of the aircraft that only the faint shiver of the fuselage gave the altitude away.

Sheng knew something was missing. There was too much calm in Nadia's departure. He ran multiple sweeps of the architecture, deeper than he'd ever gone before. David was alongside him running his own analysis on a second terminal. They checked everything, knowing that if there was hidden code, it would be discreet and buried. They went after the adjacent modules first.

Events, marketplace, and groups.

David found it hiding in the Birthday Notification Module under an account tag comment named margaret_legacy_dms.

Sheng stared at the screen for a long moment. Margaret_legacy_dms. Birthday notifications.

He said nothing. Then, he reached into a side pocket of his messenger bag and pulled out a worn birthday card. Something nagged him about that card and now the message was blatantly obvious.

Happy Birthday, Margaret.

Though thou art in heaven, we miss you still.

Remember to send reminders.

"Son of a bitch," he said quietly, a brash laugh bursting forth.

David looked over.

"What?"

Sheng slid the worn card across the desk.

David read it twice, then looked at Zhen, his eyes wide open.

Sheng nodded.

"Someone left us a map."

They searched for recursive subroutines in other modules and found none. They cut it out with the care of bomb-diffusion experts.

No alerts hit the monitor or cascading windows launched when they cut out `MARGARET_LEGACY_DMS`. The dead man's switch was quietly exorcized.

Five thousand feet and two hundred miles to the south below Nadia, a commercial United flight was banking toward the Pacific.

Gabriel Lerner sat in 24C, his knees pressed against the seat in front of him. He wasn't in a Gulfstream. He was in the normal world, surrounded by the smell of percolated coffee and the collective snore of a hundred strangers.

He watched a toddler three rows ahead struggle with a broken tablet, the child's frustration mounting as the screen remained dark. Six months ago, Gabriel would have seen this as a failure of hardware. Now, he saw it as a mercy. The child was being denied entry into the loop, saved by a dead battery. The passengers around him, the few who were awake, were bathed in blue light from their own devices, their thumbs performing the rhythmic pull-to-refresh that Nadia tuned to mimic a slot machine. They were all ghosts in their own lives, and he was carrying their exorcism in his lap.

He felt a strange, laser-focused calm. The visceral nausea from Joel's black book settled into a cold weight in his gut. He reached into his messenger bag and touched the edges of the printed scans. The half-ream of paper with the printed confessions was real.

For the first time since February, he wasn't fighting a ghost. He was going to meet the woman who knew how to use a mop to move the world, for a second time. This time the agenda was different.

He closed his eyes. He didn't see code. He saw the face of Dr. Joel Becker in the scans. A man who traded his conscience for a better AOC wine and was now waiting in the wreckage for a chance to be told he wasn't a monster. Gabriel tightened his grip on the bag. He wasn't going to California to give Becker absolution. He was going to give him a chance to survive.

In Building Seven, the graveyard shift ended.

Vira stood by the service elevator, her industrial gray jumpsuit damp with the sweat of a night's labor, her hands clasped behind her back. She didn't look tired. She appeared as a soldier might, standing watch over a captured pill box.

She knew Gabriel was coming. She knew the influencer, Suwa, lit the match. She knew the conspiracy guy had been pouring gasoline onto the bonfire. She had done her job. She'd gathered the residue, emptied the bins of the powerful, and laid the trail. Now, she simply waited for worlds to collide.

She ran a hand over the cool metal of the elevator door. Her palms were calloused, the skin toughened by decades of holding things together that wanted to fall apart. In Kharkiv, she learned that the most dangerous person in the room isn't the one with the gun; it's the one who knows where the foundation is weak.

At Argus Secure, Zhen Sheng sat in his office, watching Suwa's orchestra play.

On his primary wall-mounted monitor were fourteen red nodes. Suwa's ghost accounts were bleeding into the mainstream. The 1,400% engagement anomaly wasn't a spike anymore; it was a plateau. The public was waking up to the affinity mask.

On his desk monitor, Sheng looked at two blue dots on a flight path tracking map heading in opposite directions, the space between them growing. He'd tagged them.

G-87 (Inbound SFO). NF (Outbound JFK).

A third green dot, *V_V*, was moving north towards San Francisco.

He allowed himself a slight smile, the expression of a man seeing the architecture align. He'd preemptively drafted new contracts under Argus Security.

CHIEF TECHNOLOGY OFFICER: LERNER, G.
PRESIDENT – OPERATIONAL OVERSIGHT: TKACHENKO, V.
CO-EXECUTIVE DIRECTORS

He would redeem Prizmatik not through alignment sessions, but through evidence. He would put the Ghost in charge of the room of the men with genuine faces. He would give the subject the keys to the laboratory.

Even Becker. He would keep the doctor. He would change his equity package and put him in Nadia's old chair. Not to innovate, but to realign him to where he should have been all along. To be the Managing Morality Officer who actually had to look at the blood on the floor.

Sheng picked up his phone and sent a single, encrypted text to a number he'd never messaged before.

The Adaptive Friction Optimization overlay is gone. The air is clear. Land safely. We have a company to build.

There was one more piece of the puzzle to put in place. He did that immediately after he sent his message to Gabriel.

CONVERGENCE UNMASKED

The SFO arrivals curb was its usual chaotic slurry of travelers. Vira's 2003 Civic sat at the far end of the terminal curb like a jagged rock in a fast-moving stream. The muffler still rattled its arrhythmic, metallic cough that she explained as the way to identify her car.

Gabriel stepped out of the sliding glass doors, the damp, fog-laden Bay Area air settled over him. He inspected the line up of waiting vehicles for the car that sounded consumptive. When he saw it, he felt a strange, physical tether snap into place. He wasn't the subject anymore. He was a man with a half-ream of paper of vindication and a destination.

He walked to the passenger side. Vira didn't get out. She didn't even unlock the door until he was standing right there, his face reflected and distorted in the salt-streaked window.

He pulled the door open and sat. The interior of the Civic smelled of lemon, fake pine freshener and black tea. The scent of the virus. Vira was wearing the same navy parka she'd worn at the gas station, her hands steady on the steering wheel. She had a weathered, indestructible look of a woman who survived more than corporate scandal.

"You make good time, Mr. Brownstone" Vira said. It was full, resonant, carrying weight. She was actually smiling. Faintly, but Gabriel noticed.

"Sheng reached out to me. I mean... I know, but I don't know how he even knew about me... and how he got my number," Gabriel replied. He reached into his bag and pulled out the printed pages of Becker's book. He held them out to her, the paper crisp in his hands. They weren't shaking now.

Vira didn't take the pages. She just glanced at them, then back at the road as she pulled into the flow of traffic.

"I don't need to read sins again, Gabriel. I smelled them in trash for years. I need to know if you are ready to sign death certificate."

Gabriel looked at her profile. Her sharp aquiline nose, her wide, piercing blue eyes that never seemed to blink. He realized that while he'd been theorizing the social/digital apocalypse from his brownstone, she had been living inside all of it.

The 2003 Civic with too many miles on it merged southbound onto the 101, its fourth cylinder protesting with a discordant shudder. Gabriel shifted the half-ream of paper in his lap. He felt the vastness of the California night pressing against the glass. It felt compressed and wide open at the same time.

"Sheng said the air is clear," Gabriel said, his voice still carrying a dry quality of dehydration from the flight. "But the birthday module – it's still alive, Vira. If Nadia realizes she's lost the board, she'll pull the pin simply by not being there."

Vira didn't look at him. She kept her eyes on the taillights of a tech-shuttle in front of them. Her hands were steady on the wheel. Her appearance was that of a woman driving a tank into battle, not a rusted-out Honda onto a highway.

"She is already pulling pin," Vira said. Her voice was a low, resonant growl that seemed to vibrate in the floorboards. "She is on flight to New York. She is going to your brownstone, Gabriel."

Gabriel felt a cold spike of adrenaline. He gripped the armrest of the car, the cheap plastic crackling under his palm. His heart hammered a frantic, uneven rhythm against his ribs. His brownstone cage on East 87th Street was now a physical target.

"My home? If she gets to my terminal…"

"She think you are variable she will deploy," Vira corrected, cutting him off with a sharp, sideways glance. "She thinks you are there and broken, to be used as detonator. She does not know Sheng changed game."

Gabriel stared at her. "What does that mean? Sheng didn't tell me…"

"Sheng moves like ghost, but his actions are hammer," Vira said, her shortened English construction, a remnant from her Ukrainian life. Gabriel had grown fond of her phrasing. She reached out and tapped the dashboard three times, a sharp fingernail-click on the plastic. "She will find locked door. Then she will find men with kind faces waiting. Ones who do not care about satisfaction index."

Gabriel let out a breath he felt like he'd been holding since the Shell station days before. The tension in his shoulders melted.

"So, we're... we're done?"

Vira steered the Civic into the fast lane, the engine straining. "No. We are witnesses, remember? Sheng handles high-level treason. We go to Building Seven. We make sure Potemkin backup doesn't replace one lie with another. We are to ensure when lights come on at Prizmatik, they show floor as is. Dirty."

Silence settled between them for several moments. The muffler rattled beneath the floorboards. Outside, the dark edge of the Bay appeared and vanished between stretches of industrial parks and sleeping office campuses with darkened windows that reflected nothing back.

Gabriel looked sideways at her.

"You really want her dead," he said quietly.

Vira's eyes remained fixed on the freeway.

"Yes."

The honesty of it landed harder than rage would have.

"She has to answer for what she did," Gabriel said "...but execution isn't justice."

Vira gave a faint exhale through her nose.

"She already answer, Gabriel. Millions of times." Her grip tightened slightly on the wheel. "Every delay. Every fight. Every lonely person she turn against other lonely persons. That was her answer."

The freeway curved gently south. Headlights swept through the cabin, illuminating the pale scar near Vira's jawline before vanishing again into darkness.

Gabriel suddenly realized he knew almost nothing about her life before Prizmatik.

"That doesn't mean we become executioners," he said.

"No," Vira replied evenly. "It means someone finally becomes consequence."

The sentence settled into the cramped interior of the Civic with immense weight.

Gabriel stared out at the passing lights of San Mateo off to the right. He understood her intellectually. That wasn't what unsettled him.

What unsettled him was that part of him understood her emotionally too.

"She's still human," he said finally.

Vira was quiet for several seconds.

"You still believe human means redeemable."

Gabriel looked at her. The green glow of the dashboard sharpened the hard planes of her face. There was exhaustion there now beneath the steel. Something older than fatigue.

"And you don't?"

"I believe human means capable," she said. "Capable of kindness. Capable of horror. Depend on which part gets rewarded."

Adaptive Friction Optimization. Conflict persistence. Emotional throughput. Suddenly all of it sounded less like software architecture and more like an autopsy report for modern civilization.

Vira glanced at him briefly.

"You are good man, Gabriel. That is why Sheng picks you. And why Nadia want to use you."

For months he lived surrounded by ghosts – encrypted packets, manipulated swarms, and disembodied rage. But inside the rattling Civic that smelled faintly of lemon cleaner and tea, sitting beside a woman with calloused hands and frightening certainty, the world suddenly felt dangerously physical again.

"If we build the next system on bloodlust," Gabriel said softly, "then Nadia still wins. She just changes uniforms."

For the first time since he'd entered the car, Vira turned and really looked at him. Not tactically or analytically, but human to human.

The hum of the 101 freeway filled the silence between them.

Finally, she nodded once.

"Maybe," she said quietly. "Or maybe this is first time people like her know fear instead of quarterly bonus."

Despite everything, Gabriel laughed softly.

Vira heard it. The corner of her mouth moved upward almost imperceptibly. It changed her face completely.

Gabriel read through more of the pages.

Names. Layers. Structures. Executives. Engineers. Legal insulation wrapped in language that meant nothing and everything at once.

There were people who built the system. And people who ran it.

And a smaller, quieter group who understood it from the beginning.

One name appeared more than once. Martin Reeves.

FAILSAFE PROTOCOL

The elevator moved with smooth, noiseless ascent, a pressurized capsule rising through the dark marrow of Building Seven. Inside the metal box, the air was thick. The scent of Vira's damp parka collided with the smell of Gabriel's travel-worn clothes.

Gabriel looked at the heavy, metal, matte-black Argus card in Vira's hand. It wasn't the plastic white employee card she normally used. The Argus card bypassed security everywhere on campus. It wasn't for daily use. It was for when Sheng needed her to stop being invisible. That was why she kept it hidden.

"This has been in play far longer than you knew," Gabriel said as he took a steadying breath.

"Sheng found you, didn't he?" Gabriel asked, his voice sounding small in the confined space. "He didn't just notice a janitor. He went looking for the ghost."

Vira didn't turn. She watched the floors flicker on the small LED display: *3, 4…* She stood with her feet shoulder-width apart, maintaining the stance of someone who learned early in life that the floor could drop away at any moment. Her hands were folded in front of her, knuckles gray under the elevator's LED lights. But as the light hit the side of her face, Gabriel saw it again. That haunting, symmetrical perfection that looked like a glitch in the world of ordinary. Her beauty was a sharp, eastern European blade, the kind that made her a high-value target for the men who traded in flesh. It was a face fought over, bought, and broken, and yet there she was, standing in the heart of the most powerful social media company in the world, holding the only key that mattered.

"He is man who builds keys to all cages," she said. Her voice was steady, but there was a tremor of old earth in it. "That is what he does for living, Gabriel. He knows that the most dangerous part of any cage is the part the architect does not see but that a flea inhabits. He needed someone who knew how to live in the shadows of machinery. Someone architects never look at."

"A sleeper cell," Gabriel whispered.

"A failsafe," Vira corrected. She finally turned to look at him, and for a second, the cold Virus_V persona cracked. Her eyes were weary. There was a vulnerability there, a shadow of the girl from Kharkiv who'd witnessed the sky turn red before she knew how to read a map. "He came to apartment. I thought he was from Government. But then… he showed me card. He did not talk about engagement depth or behavior architecture. He talked about home. He talked about how a city looks right before the lights go out. He told me Prizmatik was shifting from light, building darkness he couldn't stop himself from boardroom. He asked me to be eyes and ears. To find residue algorithms could not scrub clean."

Sheng hadn't placed her inside Prizmatik. He found her there.

Three years earlier, buried in low-level system logs no one thought to audit, the pattern was already there… irregular access paths, brief, precise exposures to data that should have remained unseen, then nothing. No extraction. No flags. No noise.

None of it was accidental.

He traced it back, expecting a mistake. Instead, he found a pattern. Then he found a cleaning badge. The badge of a woman who moved through the building after hours with no declared purpose beyond her role… and yet, somehow, knew exactly where to look.

He hadn't approached her immediately. He watched. He waited.

When he finally spoke to her, it wasn't a recruitment.

It was recognition.

Gabriel felt a visceral chill. Every discovery Vira made… the shredder receipts, the notebook, the thermal paper printouts in trash bins, the incongruous threads that didn't belong… had been calculated extractions.

"He has a rescue plan for you, you know," Gabriel said, the half-ream of printout in his messenger bag feeling heavier by the second. "He must have. He told me he's clearing the air. He's obviously been tracking you as closely as he's been tracking Nadia. And… me." That realization hit him hard.

Vira's jaw tightened. For the first time in years, the soldier mask flickered. Her breath hitched; a small, jagged sound that betrayed the fear she'd spent a decade burying. She reached out and pressed the emergency stop button.

The elevator jolted, hanging in the void between floors. The silence that followed was absolute, save for the faint, electronic hum of the building's nervous system.

"I do not know about rescue plan," she said, her voice dropping to a low, dangerous register. She looked down at her hands, the hands that mopped floors and survived a life that would have pulverized lesser women. "In my country, you do not wait for rescue. You finish work. You do not trust man who says he will catch you when you fall."

Gabriel watched her. He didn't reach for her shoulder. He knew her perimeter was still electrified, no matter how much he wanted to. Instead, he stepped into her line of sight, forcing her to see him not as a target or a contact, but as a man who finally caught up.

"It's not a rescue, Vira. It's an admission," Gabriel said softly. "Sheng isn't using you. He's obviously in awe of you. Because you're the only one who actually knows what the data *costs.* In blood and bleach and security and humanity."

He paused, the half-ream of Becker's confession heavy against his chest.

"I spent a year reading you," he said, and this time his voice didn't waver. "I've spent months staring at the things you salvaged late at night - all that residue. I don't want to go back to a world where I don't know what you're seeing. I don't want to go back to the screen."

Vira looked at the emergency stop button, then back at Gabriel. The optics he'd written about years ago was finally working. He was seeing her. A small, dry smile touched the corners of her mouth, a brief bridge between the tech-developer gone rogue and the survivor.

She didn't move toward him, but the wince was gone. She reached out and pulled the button back. The elevator silently resumed its climb.

"Stay close, Gabriel," she said. "The light on the sixth floor is bright. But you might need someone who knows where exit is in dark moment."

She took a deep breath to steady the growing something, the unfamiliar.

"He is a master, Gabriel. He knows when meat is ready. But he forgets that flea in the carpet also hears what architect says when he thinks he is alone. He thinks he can align us to his next vision. He does not realize we have already seen what his visions do to the world."

"Vira… What Prizmatik became was not his vision. I'd bet a lot of money on that." Gabriel said quietly, almost to himself.

The doors slid open with a soft, expensive chime.

The 6th floor was a sea of dark glass and silent workstations, a graveyard of high-arousal metrics. At the far end, in an Executive Sanctuary reserved specifically for board members, a single amber light was burning, casting long, distorted shadows across the carpet.

"He is waiting," Vira said, stepping out into the shadows. She unzipped her blue parka, her posture returning to her rigid, indestructible line, but her eyes remained wide, taking in the luxury of the floor she had only ever seen through the bristles of a brush and the swipe of microfiber cloth.

"Now he has residue he's been cultivating for three years. Let us show him that flea has sharp teeth."

Gabriel followed, the weight of the book throbbing against his chest.

The air of the 6th Floor sanctuary was cool, smelling of sandalwood and unspoken power. And there, behind a long desk milled from a single block of aircraft aluminum, Zhen Sheng sat, waiting to offer them the world.

OVERSIGHT

The Executive Board Sanctuary on the sixth floor didn't feel like a corporate office; it felt like a chapel dedicated to governing the weight of the world. The scent of sandalwood was thick, almost tactile, masking the clinical ozone of the building's cooling systems.

Zhen Sheng did not stay behind the aircraft-aluminum desk. As the doors slid shut, he stood. Gabriel saw a man without armor. Sheng looked tired. His tailored suit was slightly rumpled, his tie loosened. He walked around the metal monolith of his desk and stopped four feet from them.

He didn't look at Gabriel, instead stepping directly to Vira.

"Vira," he said. The way he spoke her name wasn't a command; it was an acknowledgment. He bowed his head. The deep, traditional sign of respect that vibrated in the quiet air. "I know the history they tried to bury in Kharkiv. I know the type of men who thought they bought you. They were fools. They thought they were buying a commodity, but they were actually inviting a wildfire into their home."

Vira stood frozen. Her hands, still gray with the tension she'd held in the elevator, self-consciously clutching the fabric of her parka. The soldier mask was there, but beneath it, her throat moved in a hard, visible swallow. The amber light caught the unshed moisture in her eyes, making her face look fragile.

"I asked you to be my ears, my eyes, my vacuum" Sheng continued, his voice soft but resonant. "But you became the conscience of this company. While I sat in these rooms pretending to align with vipers, I knew you were in the vents. While I sat in my office at Argus or at home, I knew you were the only thing that was real in a building full of refracting mirrors."

Sheng took a step forward, his eyes never leaving hers. He straightened his posture, hands at his sides, and spoke with a startling, rhythmic clarity.

"Dyakuyu tobi, Vira. Za tvoyu sylu, za tvoyu pravdu. Ty bilshe ne prykhovana. Teper ty – osnova."

Vira's breath caught, a ragged intake of air. The sound of her mother tongue in this room of power, aircraft-aluminum and sandalwood was a physical shock, more powerful than any security breach. She stared at Zhen, seeing for the first time not the man who found her, but the man who protected her soul from the corporate machinery.

"*Dyakuyu, shcho pobachyly mene,*" she whispered, her voice thick with the iron of Kharkiv. "*I shcho poviryly v mene.*"

She lowered her head, not in subservience, but in the profound respect of a soldier acknowledging a commander who brought her home.

Sheng gestured to the wide table behind him. Two leather portfolios and a small, steaming cast-iron teapot alongside three clay cups were the only thing on it.

"Gabriel," he continued. "I know about the developers meeting in SoMa," Sheng said quietly. "The one where they floated adaptive emotional escalation as a growth mechanism. You walked out before the end. Most people in this industry negotiate with their conscience. You simply left the room. That takes an enormous amount of courage. That is why you're here and why I have deeply held respect for you."

Gabriel's internal mechanisms rearranged themselves.

Zhen continued...

"Nadia Farouk is currently being met by federal enforcement in New York." Gabriel's breath hitched, his heart performing a sudden, violent throb against his ribs. He remained silent, the half-ream of Becker's confession suddenly feeling like a holy relic in his arms.

"I am spinning off the security and ethics layer of Prizmatik into a standalone entity," Sheng said, turning back to Vira. "It is called Oversight. It will have total, unvetted access to every packet, every injection port, and every ghost variable in the Prizmatik ecosystem. It will not report to a Board. It will publicly report the truth."

He stepped closer, his eyes locked on Vira's.

"I am offering you the position of Co-Director of Operations. With Gabriel as the other Co-Director. Equal authority. Total autonomy. A budget that ensures you never have to see the shadow of a mop again. Unless of course, you choose to."

Vira's soldier-straight posture finally buckled. It wasn't a collapse; it was a surrender to the reality of being seen. She reached out, her hand trembling as she grasped Gabriel's arm to anchor herself to the Earth. A single tear tracked down her cheek, leaving a clean line.

"You... you give keys to a flea?" she whispered, her voice cracking.

"The flea is the only one who knows where the blood is hidden," Sheng said. He reached out and placed a hand on her shoulder. A brief, grounding touch. "Your days of living in the shadows are over, Vira. I don't want you hiding in the carpet and the vents anymore. I want you in the light. I need you in the rooms that Nadia used to run, putting fear of being seen into the hearts of those who would use people to their own benefit."

Gabriel stood there, paralyzed with awe. He swiveled his gaze from Sheng to Vira, and back again, the silence of the room ringing in his ears. Zhen had built a fortress for people the world tried to break. He clutched the half-ream of paper in his hands, then looked at Vira, whose face was now tilted toward the amber light, stripped of the virus and the janitor and the flea, leaving only the brilliant woman who'd been through so much.

Sheng turned and walked back to the teapot and poured three small ceramic cups. The steam rose in the sanctuary, a quiet, ancient signal and ritual of peace.

"I have already leveraged the Prizmatik board. They've acquiesced to my demands. Reeves is in the next room with paperwork," Sheng said. "The contracts are Argus-adapted. But before we sign, we drink. We drink to the residue."

Gabriel finally found his voice. It was a rasp, a shadow of the leader of the Static he used to be. He turned and gazed upon Vira.

"To the flea," Gabriel whispered.

Vira looked at him, the small smile from the elevator returning, but this time it reached her eyes. She realized she was still holding his arm.

She picked up the small ceramic cup, her fingers curling around the warmth.

"No," she said softly. "To prisms scattering light."

Sheng methodically set his cup down on the aluminum surface. It never made a sound yet closed the door on the emotional vulnerability of the last half hour. He reached into the breast pocket of his blazer and pulled out a small, encrypted glass drive, the kind Argus developed and used for high-level tactical wipes.

"I am a man of my word," Sheng said, his eyes shifting between Gabriel and Vira. "But I am also a man who knows that a clean floor does not stay clean without a guard at the door.

And I believe you are both thinking the same thing: *What is the catch?*"

Gabriel adjusted his glasses, his voice finally regaining resonance. "The catch is Prizmatik. You're asking us to save a machine that evolved to amplify, then harvest rage. You can't just realign a tumor, Zhen. You have to excise it."

Sheng smiled, and for the first time, it wasn't the smile of a master chef. It was the smile of a general.

"Exactly," Sheng said. He tapped the glass drive. "This is the Potemkin kill-code. Within forty-eight hours, every BMI-weighted algorithm, every conflict-initiated injection, and every affinity mask will be purged. Not hidden. Not archived. Deleted. The servers will not be realigned. They will be scrubbed. Only then will we rebuild."

Vira leaned forward, her eyes narrowing. The soldier was back, but she was no longer in the shadows. She reached out and touched the glass drive, her thumb tracing the encryption seal. She wasn't looking at the tech; she was looking at the power it represented.

"And all users?" Vira asked. "The millions who have been fed vile woman's poison? You cannot delete memory of riot in Ohio."

"No," Sheng said. "But you can show them the footage of the projectionist. Oversight will not just watch the company. It will publish weekly reports to the public. Tuesday, at 9:00 AM, every week, the world will see exactly what the algorithm tried to do to them the week before, until trust is regained. Transparency is the only disinfectant that doesn't leave a film. I'm calling it The Residue Report."

Gabriel felt a sudden, electric surge of intense clarity. This wasn't just a job. It was a counterinsurgency. Sheng was turning Suwa's Orchestra into a public audit.

"You're going to let us post the truth," Gabriel whispered. "Even if it kills the IPO. Even if it kills the value of what you built?"

"The stock valuation is a ghost," Sheng said dismissively. "I am interested in the foundation. And I am interested in the two people who were smart enough to survive it. And..." he actually chuckled, "once the truth is out and the public understands, with a little hand holding and education? Prizmatik will be worth even more. Obscenely more."

He turned to a wall-mounted monitor and motioned them over. It was one that wasn't connected to the Prizmatik grid. Sheng had it wired to its own isolated external feed long ago. Currently, it was showing a low-light feed of a block in Manhattan from the corner. East 87th Street. Gabriel was looking at his house.

"One more thing," Sheng said, his voice clear and concise. "I told you I was clearing the air. Nadia Farouk's flight landed an hour and twenty minutes ago. She is approaching your home, Gabriel. She is emotionally carrying a Dead-Man's Switch she thinks is still active and available to her. She is expecting a subject. She is expecting *you*. She is expecting the breakdown we simulated with Kael's script."

"Wait... You knew about Kael's script?"

Sheng smiled again and laughed. "I helped him write it."

Gabriel gripped the edge of the table, his knuckles turning white. The depths of Sheng's planning. Then the physical reality of the threat hit him. The distance between the sanctuary in California and his home in New York suddenly felt terrifyingly short.

“The FTC,” Gabriel said, remembering the “Men with kind faces” Sheng mentioned and Vera lived through.

“They are in the foyer,” Sheng confirmed. “But they are not the ones who will end this. You are.”

Sheng pushed a separate tablet toward Gabriel. It was a remote uplink to the East 87th Street terminal with a separate hack into his front door intercom.

“She wants to talk to G-87,” Sheng said. “She wants to see the Subject bleed before she pulls the pin. She’s timed this all so that her absence would automatically trigger the switch as she watches you fall. May I suggest… you give her a different kind of satisfaction.”

Gabriel looked at Vira. She nodded once… a sharp, military gesture of permission. She reached over and placed her hand on his. It was the same rough hand that had saved him. It was a silent transfer of strength. The flea and the leader were no longer variables. They were the architects now.

Gabriel reached for the tablet. His fingers were steady. The ring light blindness was gone. The voice was gone.

"What do we do?"

Vira was silent for a long moment. She stared into Gabriel’s eyes, and simply said "we tell truth."

He took a deep breath as the weight of what he was about to do landed squarely upon his shoulders. Without breaking eye contact with the Ghost, he said “open the channel please. Let’s show her what a perfect Satisfaction Index looks like.”

EPILOGUE: ACCUMULATED CONSEQUENCES

The fog rolled in through the Golden Gate and over the Presidio, but from the terrace of the house on Pacific Avenue, it didn't look like a shroud. It looked like a cleansing microfiber blanket.

Zhen Sheng had seen the early days, the optimism, and the belief that the network would elevate discourse. Not everyone shared that belief. Some saw it as a tool for what it would become. A system.

Now, watching the world through the lens of Oversight, he saw something better than optimism: he saw honesty. Oversight's headquarters in a Mission District warehouse didn't feel like a corporate bunker or a glass and chrome castle; it felt like a comfortable workshop. The walls were covered in actual whiteboards – the messy, erasable kind – where engineers debated humane and social ethics with the same ferocity their predecessors once reserved for latency. There were no meditation or sleep pods, digital walls or living walls or frictionless surfaces. Instead, there was the smell of strong coffee, the sound of laughter, the occasional pops of ping pong balls being smashed back and forth with enthusiasm... and a collective sense of purpose that didn't require a dopamine loop to sustain.

Jimi's *Electric Ladyland* blasted through Zhen Sheng's listening room at a volume approaching a live performance. Sheng stood between his McIntosh monoblocks and the B&W 801 D4 Diamond Series speakers, meditating upon the orange glow of the tubes. He flashed back to a story he'd heard years ago... about the man in 1993 watching a video feed of goldfish... and the client who understood, before anyone else in the room, that the conversation was never the point.

He waited for the specific, cosmic swell of "1983 (A Merman I Should Turn to Be)." When the line arrived, Sheng didn't just listen; he sang along to the dismissal of the "giant machine" and the "leftover energy" it once fed upon. He sang about the peace of mind found by walking away from the machinery and into the sea. He sang along badly, his voice a rasp against the high-fidelity perfection of the system and the room, and he laughed. It was the laugh of a man who had broken the prism and found the truth hidden in the light.

Martin Reeves sat alone on the terrace of his penthouse watching the first light of dawn spread across the Bay from behind Mt. Diablo.

At seventy-four, his days belonged entirely to him. No board meetings. No investors. No earnings calls. Only the city, the water, and that particular quality of early light that made San Francisco feel like a promise being renewed.

He had not slept well. He rarely did anymore.

A strong latte steamed beside him. He had been sitting with his phone in his hands for twenty minutes, not scrolling exactly, but moving through the feed, not looking for anything, while unable to stop looking.

He watched what Prizmatik became. He watched from a distance that felt safe and wasn't. He told himself for years that the original sin wasn't his. That he only enabled the possibility. That what others had done with it was their reckoning, not his.

He no longer entirely believed that.

Then a support group appeared in his feed.

A hundred thousand members. Patients. Families. Caregivers. The occasional doctor offered what he or she could. Strangers helping strangers navigate cancer treatments. Questions answered at three in the morning by people who owed the asker nothing. Meals delivered across cities to people they'd never met. Friendships formed in the specific crucible of fear that only people inside of it could fully understand.

The group was named PrairieMoon.

Martin sat very still.

His wife spent her final year searching for exactly this. Not a cure. Not a miracle. Just people who understood what she was living with. Who would sit awake through the night with her, virtually holding each other's hands through the hardened darkness that arrives at 3 AM when the body no longer cooperates and the mind won't rest.

He built the room but she arrived too early for the door to open.

He gazed toward the sunrise for a long time. The tears came without drama, the way they did now – not grief, but something quieter and more complicated. The feelings of a man who spent thirty years wondering whether he'd handed the world a gift or a weapon and finally lived long enough to understand that the answer was yes. Both. Completely. Irrevocably.

The sun finally cleared the horizon, scattering gold across the water. The Bay Bridge filled with the first commuters of the day, people beginning ordinary days inside a world he helped make and couldn't entirely account for.

He picked up the latte. It was cold. He drank it anyway.

Then he whispered the same words he had said to an empty office in Santa Clara, watching goldfish swim in a tank he would never visit, on a Tuesday night in 1993 when the future had felt like nothing but possibility.

"I'll be damned."

This time he knew exactly what he was damning.

He sat with that, in that early light, for a long time.

Gabriel sat in a deep, thick mocha-hued leather chair that didn't ergonomically cradle his spine or track his torso stress. He was in a living room. He was holding a book. "We" by Zamyatin. Thick, cream-colored pages that smelled of ink and old paper. Beside him, a glass of sparkling water sat on a coaster. No sensors or smart watch. No HUD. No 6K display. He didn't need to see the world through a filter anymore; he was the one who kept the lens clean. And he enjoyed his janitorial role immensely.

His hands were steady. The phantom voice that had haunted him for a decade vanished the day the FTC led Nadia Farouk down the steps of his East 87th Street brownstone in handcuffs. The half-ream of

Becker's confession wasn't in a messenger bag anymore. It was in a glass case in the entry of Oversight, Inc., a memento mori for every engineer who walked through the door. It was required reading by every new employee before they signed their hiring papers. Those separate signature pages were in another glass case next to the printout.

A door creaked open behind him. Vira stepped out, wearing a long, soft Alpaca sweater that reached mid-thigh. It matched the gray of the Pacific fog. She didn't look like a Ghost or a failsafe anymore. The blade of her face softened by time and care, the sharp edges of survival replaced by the elegant movements and gravity of a woman who owned the ground she walked on.

She leaned against the back of his chair, her hands still scarred, though no longer rough, a permanent badge of her history, alit on his shoulder. She looked up toward the next block, where the roofline of Zhen Sheng's house peeked through some unruly Cypress trees. They were neighbors now. The Architect, the Subject and the Virus, living in the light they fought to preserve.

"Kyle just texted me. The audit for our London hub just cleared," Vira said. Her accent was still there, a grounded anchor in the air. "They tried a 100ms throttle on some dissent tags. We caught it in under four seconds. The fine will hit their quarterly before the sun sets in the UK."

"Good," Gabriel said, not looking up from his book. "Now that we've lowered the evidentiary threshold, even 50ms gets tagged. Let them pay for the audacity."

Oversight was no longer a startup; it was the world's honesty machine and it oversaw every social media platform across the globe. It was now the flea with a global megaphone and cross-border authority. Under their direction, the world wide web hadn't become a utopia. It simply became a place where the projectionist could no longer hide.

Gabriel pulled Vira onto his lap and kissed her as she laughed and tried to squirm away. She didn't try too hard as they shared breath.

"I doubled the order for the Zabar's shipment for next week. You've been eating more lox darling..." she whispered in his ear as she lovingly played with an errant curl of his hair.

In a Federal Correctional Institution in Danbury, Nadia Farouk sat in a room that provided no aesthetic filters. There were no triple-monitor arrays, no Aeron chairs, and no engagement metrics. There was only the gray concrete of the yard and the 15 years remaining on a 20-year sentence for Seditious Algorithm Manipulation. The charges were many and severe. Many were dropped in exchange for cooperation. She gave them as much as her ego would allow. It wasn't enough to overcome the mountain of residue, the physical evidence that Vira gathered and hidden in a locked janitor's closet on campus. It was all introduced at trial.

But the world remembered her. She was the face in every ethics textbook, the poster child for the era of Silicon Valley manipulation, hubris and cruelty. She wanted and needed to be a goddess of data; instead, she became the cautionary tale of moral hubris, a specter trapped in a cage she hadn't built.

In Seoul, a woman named Soo-Ah stood in the back of a high-end boutique, folded silk blouses with practiced precision. The lights were gone. The Bisou Spike was a faded memory in a digital archive. Occasionally, a customer would look at her, at that symmetrical, beautiful face, and wonder if they'd seen her before.

She'd moved back to her home country after the scandal broke very publicly online. Over 70% of her subscribers abandoned her platform within two weeks of her very public dismantling of the Prizmatik outrage design. The offers for review and endorsement rapidly dropped, then stopped coming altogether. Her orchestra moved on to a different conductor who didn't publicly excoriate their corporate sponsors.

She folded the blouse with practiced precision, aligning the seams the way she used to align a frame.

No one was watching.

For the first time in years… that felt like the point.

In a windowless basement in Seattle, the late-night posts ceased.

Kael was intrigued by a series of posts on Prizmatik and an obscure rage-inflated bulletin board. He wrote an automated script to track @SLCNMSTER69 and sent Zhen a heads-up that there were people out there in the broader world that were seeing the truth. It was the very truth that they all fought for, but that @SLCNMSTER69 actually exposed to the world.

It was Kael and Zhen's script that noticed Herbert's inactivity. It had flagged the stagnant node that had been consistently active for years, telling truths no one cared about. Now it was an IP address that hadn't moved a cursor in seventy-two hours. Then a hundred hours.

His mother also noticed. The tap tap tap striking the keyboard in the basement in the middle of the night was silent. Then, pizza and soda deliveries stopped.

The smell hit her first: must and mold, the rotting food and the rotting body, a smell she didn't know, though instinctively did. She didn't go downstairs.

When the local authorities arrived, Herbert was in his gaming throne, his obese body slumped down. His eyes open, staring at a screen that had timed out into a black void. His right hand was curled into a claw, still clutching an empty, crumpled bottle of Mountain Dew. He died in the middle of a thread, a data point that flatlined in the dark, Han Solo, the only witness to the last moments of Herbert's life.

Back in Presidio Heights, the sound of laughter drifted through the house. Two children — a boy with Gabriel's thick glasses and a girl with Vira's wide blue eyes and indestructible jaw — chased each other about the hallway upstairs.

Gabriel dropped his book to the floor as he held onto something incredibly precious in his lap. He gazed at Vira, his wife.

"Zhen is coming over for champagne and crab at five," he said.

Vira smiled. A real, un-simulated expression that ignited her eyes. She reached up and cupped his boyish face, her palms lifting it, demanding eye to eye attention.

"The champagne is chilled, сонечко," her accent a grounded anchor in the evening air.

Gabriel looked at her, seeing neither the virus or the janitor, but the woman who had taught him that visibility is a form of love. "And the floor?" he asked, pulling her closer.

Vira laughed, a sound more resonant than any engagement metric. "Floor is clean, Gabriel. Is *very* clean" as she kissed him with an honest ferocity.

The doorbell rang.

Gabriel exhaled softly against her forehead. "That'll be Zhen."

"Or the lox delivery for you my kokhányi," as Vira smiled into his eyes.

Gabriel laughed as she slid off his lap. He rose, crossed the room barefoot to the foyer and opened the front door.

Joel Becker stood there holding two bottles by their necks like offerings.

"I was told there would be crab," Becker said.

The labels caught the porch light as Gabriel stared down at them. 2005 Romanée-Conti.

"Jesus Christ, Joel… you didn't have to bring *those*!" Gabriel gasped.

Becker shrugged lightly. "Zhen insisted I stop bringing Russian River Pinot to dinner."

They laughed.

Gabriel looked at him. Really looked at him. Not at the compliance officer. Not at the man who wrote his confession. Not at the exhausted architect of guardrails no one wanted installed.

Just Joel.

"Please… come in," Gabriel said warmly.

Becker stepped inside as Vira approached from the living room.

He flashed back to the night he had watched her studying his monitors. He'd recognized the look, the intentness of her gaze, though she did everything to shield herself from discovery. She'd possessed the look of someone understanding exactly what they were looking at. And then, her intensity that she hid so well as she feigned ignorance in Ukrainian, had laid the path for everything that followed.

His intuition had been right. He was also ashamed that was so wrong about himself at the time. He looked fixedly at Vira approaching and saw the bravest person he'd ever met.

Gabriel took the wine as Vira took Joel's hands, smiled and kissed him lightly on the cheek.

"Zhen invite you?" she asked.

"Apparently I'm no longer a threat to corporate governance," he laughed, glad that so many demons from their collective past had been exorcised.

"We're still monitoring you," Gabriel chuckled as he strode towards the butler's pantry with $40k of wine in his hands.

"Good," Becker answered. "That means the system is working as intended."

Zhen strolled in just as they went to the butler's pantry to decant the wine.

“Good” he simply said, as he clasped Joel’s hands in his with genuine fondness. He knew he’d been right about Dr. Becker all those years ago when he’d sewn the seeds for his hiring through multiple avenues of recommendation and teasing his name throughout Marcus and David’s reign.

Vira stepped out the rear French doors and stood alone on the back deck as the others drifted toward the kitchen.

The evening air carried the perfume of damp grass, eucalyptus and the saline breath of the Pacific. Somewhere beyond the low wall of the Presidio, a coyote called into the darkness.

She didn't triangulate the distance or direction of the sound. She didn't formulate the exits and angles of escape that had once been as automatic as breathing.

It simply was a coyote, calling because it wanted to. Another answered, farther away. It didn’t trigger anything in her except admiration for lives lived in freedom.

She stood with her hands loose at her sides, feeling the chill come in off the water. Through the lit windows behind her, she could hear Gabriel saying something that made Zhen laugh – a real laugh, unguarded, the kind that didn't know it was being heard.

A door somewhere in the house slammed.

She felt nothing.

Not the old contraction. Not the half-second inventory of who was behind her and how far she was from the street. Just the sound of a door, in her house, where people she trusted were making dinner and opening wine as her children played upstairs.

She stood there a little longer, in no particular hurry to go anywhere.

The coyote called out once more and went quiet.

Vira turned and walked back toward the light.

The fog bank broke over Menlo Park, and for the first time in a long time in the history of Prizmatik, as the sun hit the glass windows of Building Seven, it didn't reflect back. It went right through, scattering its purity into a thousand hues of truth.

GLOSSARY

The following terms and protocols are derived from the original UCLA Boelter Hall server logs (2003–2005) and the subsequent hardened architecture developed during the SoMa expansion.

- **0x8F2A1C9E:** This is a hexadecimal memory address. In a system log, it represents a specific place where data lives. To an engineer like David, seeing an error at this address is a deep root clue that something is amiss.
- **Affinity Mask**: A content suppression protocol that deployed high-joy, high-engagement materials: animal videos, babies, reunion clips, celebration posts. Purpose: to algorithmically smother Conflict-Resultant spikes before they became visible to outside observers.
- **AFO (Adaptive Friction Optimization):** An internal Prizmatik behavioral architecture framework originally derived from the early 400ms latency experiments. AFO dynamically modulates micro-delays, interface resistance, and response timing based on real-time emotional and behavioral analytics in order to increase user engagement, conflict persistence, and platform retention. Officially categorized as an "engagement stabilization" system. AFO was capable of amplifying agitation, compulsive interaction patterns, and predictive conflict escalation at an individualized level.
- **Air-Gapping:** A security measure that involves isolating a computer or network from the internet. Zhen Sheng insisted on an air-gapped terminal for the platform's root keys to prevent remote hijacking. Gabriel used an air-gapped computer to archive and consolidate data and group discussions.
- **Algorithm:** A sequence of instructions for a computer to solve a problem. In Prizmatik, one algorithm's job shifted from "find people you know" to "predict who and what you want to see next."
- **API (Application Programming Interface):** A set of rules that allows one piece of software to talk to another. David used the Classmates API to integrate their entire data structure into the Prizmatik databases.

- **Behavioral Modification Index (BMI):** A Prizmatik metric designed to measure how much a user's real-world behavior changes after using the platform.
- **Bio-sync**: The real-time synchronization of a user's physiological data. Heart rate, galvanic skin response, sleep cycle, movement patterns, as it pertains to Prizmatik's content delivery system. Aggregated through wearable devices and mobile sensors from phone apps, bio-sync data allowed the platform to identify optimal windows of emotional receptivity and time the delivery of high-arousal content. Users understood bio-sync as a wellness feature. The platform used it as a targeting system.
- **Checksum:** A small block of data derived from another block of digital data for the purpose of detecting errors. In the Prizmatik architecture, a checksum is the system's way of saying, *"I am exactly what I claim to be."*
- **Conflict-Initiated vs. Conflict-Resultant**: The distinction Joel Becker raised in Nadia Farouk's quarterly presentation. She found the differentiation inconvenient because it was correct. Conflict-Initiated content was designed to provoke. Conflict-Resultant content was what happened to people after provocation. It ties directly into the BMI.
- **Convergence:** The moment the digital network and the human social structure become indistinguishable from each other.
- **Database Sharding:** The process of breaking a very large database into smaller, faster, more manageable chunks (shards). As Prizmatik hit 10 million users, David had to shard the Relationship Table to prevent the system from collapsing under its own weight.
- **DAU/MAU (Daily Active Users / Monthly Active Users):** The Stickiness ratio. A ratio above 50% means the app is no longer a tool; it is a vital part of someone's daily life.
- **Dead-Man's Switch:** A specific script within an algorithm. If Nadia did not check in with the system every 24 hours, the tool would automatically broadcast the decrypted Farouk Protocols to *@*.* address in the database, returning the story to where it began.
- **Digital Gaslighting**: The practice of serving a user a parallel version of the platform, a simulation designed to confirm their existing beliefs about the system's behavior while concealing the system's actual behavior. The user believes they are seeing the actual system. Instead, they are seeing a

reflection of their own suspicion, curated and amplified until it becomes indistinguishable from evidence.

- **Dopamine Loop:** A neurological cycle of "Activity → Reward → Craving." Nadia tuned the notification sounds and "pull-to-refresh" haptics to mimic the mechanics of a Las Vegas slot machine.
- **Engagement:** A catch-all term for any user interaction (likes, clicks, shares, scrolls). It is the currency of the modern web.
- **Entropy (Systemic):** In information theory, a measure of randomness or disorder. Zhen Sheng watched the Entropy Scale of the user feed; the higher the entropy, the more the algorithm had to work to force a pattern of engagement.
- **Friction:** The holy grail of systems design is to eliminate friction. The goal of making an action (like finding an old friend) so easy that the user does it without thinking.
- **Frictionless Radicalization:** is the process by which a user moves from casual engagement to extremity without ever experiencing the resistance that might have caused them to stop and reconsider. The platform removed the friction. The radicalization followed naturally.
- **Hardened Linux Kernel:** A highly modified version of the Linux operating system, stripping away non-essential functions, leaving only raw power. To harden a kernel is to close every digital window ensuring that only authorized system calls can be executed.
- **Heat Map:** A visual overlay on the HUD that shows where users are clustering. The warmer the color, the higher the obsession.
- **High-Arousal Negative Affect (HANA):** The emotional state Nadia designed and the platform learned to manufacture at scale. HANA is the condition that precedes object and resolution both, the raw physiological arousal of a system that has been told something is wrong but hasn't been told what. In this state, the user is maximally receptive to whatever the feed offers next.
- **HTML (HyperText Markup Language):** The markup language used to structure World Wide Web pages, interpreted by web browsers such as Mosaic, Netscape Navigator, Google Chrome, and Microsoft Edge to render text, images, links, and other page elements visually.
- **HTTP (Hypertext Transfer Protocol):** The foundational communication protocol of the World Wide Web. Developed by Tim Berners-Lee at CERN between 1989 and 1991

alongside HTML and the URL addressing system, HTTP allowed hypertext documents to be requested and transmitted across the existing Internet.

- **HUD:** The Heads-Up-Display administrative interface overlay that allows David and Marcus to see the network's health.
- **Injection Point:** A specific line of code where a secondary process is injected into a primary one. Nadia's engine was injected at Memory Address 0x8F2A1C9E, allowing her to override David's Bulletin Board logic with her Modification logic.
- **Intermittent Reinforcement:** A psychological tactic of giving rewards at random intervals. This is the most addictive form of conditioning known to science; it ensures a user never walks away because the big hit could be the very next click. It is used extensively in gaming logic.
- **IP Address (Internet Protocol Address):** A numerical address assigned to a device on a network so data can be routed to and from it. An IP address may identify a device on a private local network, such as 192.168.68.1 (an internal network address for a router or printer), or a public address on the wider Internet.
- **Kernel-Level Auditing:** A background process that watches the watcher. If Nadia's behavioral engine tried to access a memory address it wasn't assigned (like the private message encryption keys), the Hardened Kernel would shut down the node before the data could be compromised.
- **KPI (Key Performance Indicator):** The specific numbers that tell the board if Prizmatik is winning. For Zhen Sheng, the most important KPI wasn't just "User Growth," but "Retention," or how hard it was for a user to leave.
- **Latency:** The measurable elapsed time between an action and the system's response. David fought to minimize latency because users perceive delay as poor performance. Nadia later weaponized latency by introducing carefully calibrated response delays that increased anticipation, emotional arousal, and engagement.
- **Load Spikes:** Sudden surges in network traffic that can crash servers.
- **Lotus 1-2-3**: A desktop computer spreadsheet application released by Lotus Development Corporation in 1983,

dominant through the late 1980s, and effectively obsolete by the mid-1990s when Microsoft Excel was released. Herbert Jones used it exclusively and without apology for his investigative documentation.

- **Micro-Nudge:** Subliminal changes in the feed. A slightly larger font for a provocative post, a three-millisecond delay in showing a dislike. Designed to steer a user's mood toward agitation.
- **Multi-Layered Encryption:** Vira's onion tool wrapped every piece of sensitive data in multiple layers of encryption. To see the core (the truth about the Behavioral Modification Index), a whistle-blower would have to peel back each layer using keys held by different, disconnected servers.
- **Packet Sniffing:** The act of intercepting data as it travels across a network. While used by Gerald at UCLA for maintenance, Nadia Farouk adapted the concept into Behavioral Sniffing, intercepting the *timing* of clicks rather than the content.
- **Potemkin Protocol**: The emergency content architecture activated in the hours following Nadia Farouk's termination. Named, without irony, by Arthur Vance of The Static, after the Russian practice of constructing false fronts to conceal the truth from passing observers.
- **Predictive Modeling:** Using historical data to forecast future behavior. Nadia's models could predict with 94% accuracy which users would engage in an argument before the users themselves felt angry.
- **Port (e.g., Port 23, 80, 443):** Virtual doors where data enters a server. David's Boelter Hall builds often hid traffic on obscure ports to avoid UCLA's campus-wide firewalls.
- **Reconnection Event**: In the platform's original architecture, a Reconnection Event was the moment the system fulfilled its founding purpose: to eliminate friction. Marcus Teller kept a library of them. He read it when he needed to remember what Prizmatik was for. Under Nadia Farouk's Behavioral Architecture, the term migrated quietly into the system logs as a metric. One data point among thousands, its weight identical to a liked photograph or a shared outrage thread.
- **Residue**: What gets discarded and left next to shredder bins, in trash cans, printouts from restricted terminals. The physical trail left by institutions that have learned to be invisible to their own infrastructure.

- **Sentiment Analysis:** A machine-enabled process that "reads" the emotional tone of every post. Nadia used this to cluster angry users together, knowing that outrage is the most effective driver of engagement.
- **Server Racks:** The physical machine bodies of Prizmatik housed in the Menlo Park data centers. By 2007, the heat generated by these racks required a dedicated industrial cooling system and might be seen as a physical manifestation of the platform's upwards-spiraling growth.
- **Social Graph:** A mapping of everyone on the platform and how they are related. David built the graph to help people find each other; Nadia used the graph to see who was load-bearing, which users, if manipulated, would cause a Cascade Effect of anxiety across their entire social circle.
- **Static**: An encrypted peer-to-peer research collective founded by Gabriel Lerner. The Static spent fourteen months building what its members believed was the most rigorous independent documentation of Prizmatik's behavioral architecture in existence. They were wrong about the independence. The platform had been observing their work for six of those fourteen months, using it to calibrate the simulation it was feeding them in return.
- **System Status: Performing as Intended:** The most dangerous phrase in the Prizmatik logs. It implies that the intended outcome is a good thing.
- **Technical Debt:** The implied cost of additional rework caused by choosing an easy solution now instead of a better approach that would take longer. David's integration of the Classmates database was a massive debt he intended to pay, but Nadia used the cracks in that debt to hide her modification code.
- **Traffic Camouflage:** A method of sending ghost data through the network. To an outside observer (like a government monitor), the network looks like it's just sending cat videos and birthday invites, while the actual Onion data is hidden inside the noise of the traffic.
- **UI/UX:** User Interface (what it looks like) vs. User Experience (how it feels). Marcus knew that a perfect UI could hide a dangerous UX.

DISCUSSION QUESTIONS

1. Martin Reeves experiences the birth of the World Wide Web as a moment of profound optimism and possibility. Were those hopes naïve, or did the internet become something other than what its early pioneers imagined?
2. Several characters realize that Prizmatik is not merely reflecting human behavior but actively shaping it. At what point does influence become manipulation?
3. Herbert Jones is often abrasive, socially difficult, and easy to dismiss. Why do societies frequently ignore uncomfortable truths when delivered by imperfect messengers?
4. Vira Tkachenko repeatedly observes that invisibility can be a form of power. In what ways do modern institutions overlook the very people who may best understand them?
5. Dr. Joel Becker and Nadia Farouk approach human behavior from different perspectives. Is there an ethical limit to using technology to understand, predict, and influence human behavior?
6. Prizmatik amplifies outrage because outrage increases engagement which generates revenue. Do social media platforms create division, or do they expose divisions that already exist?
7. The novel suggests that algorithms learn from human behavior rather than imposing behavior entirely on their own. To what extent are users responsible for the systems they collectively train?
8. Several characters sacrifice comfort, careers, and personal safety after recognizing what Prizmatik has become. What obligations do individuals have when they discover that the systems they depend upon are causing harm?
9. Throughout the novel, communication technology is portrayed as both liberating and dangerous. Is technological progress inherently neutral, or do technologies inevitably shape the societies that create and adopt them?
10. The novel presents a world in which attention has become one of our most valuable commodities. How has the competition for human attention altered relationships, politics, education, and our understanding of ourselves?
11. SOAPBOX ends without offering simple solutions. Did you find the ending hopeful, pessimistic, or something more complicated? Why?
12. One of the novel's central questions is: "Who exactly is shaping whom?" After finishing the book, how would you answer that question?

ACKNOWLEDGEMENTS

I would like to thank the following for their incredible and ongoing help in so many ways: All of my Cybernaut friends. Michael Kwun for his impeccable editing skills and advice. Tammy Nelson for her ongoing encouragement. Sanji Buramindi for his expertise in coding and network systems. Keith Canfield for his always candid feedback.

I would also like to thank Tim Berners-Lee for his incredible foresight in creating the world of http/html. Without his brilliance, we would not have the amazing resources at our fingertips that we do today. Wikipedia alone is a gem and one of the more incredible endeavors in human history.

ABOUT THE AUTHOR

PRIZMATIK SYSTEM LOG — 26:28:07.
USER: LAURENCE
CONTENT TYPE: META-HUMOR
SELF-AWARENESS INDEX: DANGEROUSLY HIGH
ENGAGEMENT PATTERN: OPTIMAL
RECOMMENDED ACTION: CONTINUE OBSERVATION

Laurence Davidson is a writer, traveler, musician, sommelier, and enthusiastic observer of human behavior, particularly those parts humans prefer nobody observed.

His nonfiction debut, **The Flow: A Hedonist's Guide to Buddhism, Meditation and Travels in Thailand,** blends travel, humor, meditation, and life-on-the-road in the search for a damned good bowl of something delicious and clarity beyond the relentless noise of modern life. There are Buddhist monks. There are moments of insight. There may also be a martini or two along the journey. Enlightenment is a process after all.

His fiction turns that same suspicious and sarcastic eye toward the ways people behave when their convictions collide with ambition, power, money, ego, and the inconvenient arrival of consequences. His novel **Kill All the Butterflies** examines power, intellectual arrogance, and moral compromise through the unraveling of a man who spent a lifetime believing he understands the world.

Davidson's work circles a persistent question: not what people say they believe, but what they actually do when the stakes become real, and nobody is grading them on their behavior.

He lives and writes in Washington State's beautiful, wine-soaked Walla Walla Valley, where vineyards stretch across rolling hills, excellent bottles are dangerously accessible, and calling wine "research" remains one of the considerable advantages of being both a writer and a sommelier.

His books are available through most major retailers in eBook and paperback, with hardcover editions available when the publishing gods, production economics, and general alignment of the planets permit.

To learn more, visit **laurencedavidson.com**. For essays, observations, occasional cultural poking with a stick, and whatever else has wandered through his head demanding to be written down, visit **larryistrippin.substack.com**.

And finally, remember... it is human connection and compassion that makes us real. Not our avatars, follower counts, algorithms, crafted online personas, or those little glowing rectangles we carry around demanding our attention. That's all extraneous bullshit.

www.ingramcontent.com/pod-product-compliance
Lightning Source LLC
LaVergne TN
LVHW010714061126
851857LV00014B/1181